CW00326424

Isaac Physics Skills

How to Solve Physics Problems on Isaac Physics & Beyond!

Michael Conterio
Cavendish Laboratory, University of Cambridge

Periphyseos Press
Cambridge, UK.

Periphyseos Press
Cambridge

Cavendish Laboratory
J. J. Thomson Avenue, Cambridge CB3 0HE, UK

Published in the United Kingdom by Periphyseos Press, Cambridge
www.periphyseos.org.uk

How to solve Physics Problems
© M.J.C Conterio 2019
Content in *How to solve Physics Problems* is licensed under
Attribution-ShareAlike 4.0 International (CC BY-SA 4.0)

First published 2019
Printed and bound in the UK by Short Run Press Limited, Exeter.

Typeset in LaTeX

A catalogue record for this publication is available from the British Library

ISBN 978-0-9572873-8-9 Paperback

Acknowledgements

We are the sum total of our experiences, and so this book is the sum total of my teachers. Some of them will be found in my approach to physics, some of them in my approach to teaching, and some of them in the life that led me here. Thank you all.

More recently, I wouldn't have been able to do this without all of my colleagues at Isaac Physics. Thanks should go to Mark Warner and Lisa Jardine-Wright for allowing me to take this idea and run with it. Thanks to Robin Hughes for being an invaluable officemate, as well as running his keen eye over of the initial draft of this book. Also, thanks to fellow content writer Nicki Humphrey-Baker - it's been a pleasure working with you. To all the rest of the team - thanks for a few great years working on the project. I miss it already.

To the authors of the previous Isaac Physics books, thanks for showing me how it's done. The typesetters of those books, likewise - I have liberally borrowed from your work.

So many thanks and so much gratitude to my parents, my first and best teachers.

And of course, to my wife Jasmin, for putting up with me disappearing off to write this book. You make my life amazing in so many ways.

Contents

Introduction

You have some values. They go in equations. Sometimes you have to rearrange the equations. Sometimes you don't. Learn the equations, learn how to put the numbers in, learn a few facts about the topic and you're done. Right?

Until you come across a situation that's a bit more complicated than that. Maybe none of the equations you know quite fit, or you've got two different accelerations and don't know which to use. Whether it's a question that's been set to you by your teacher or one that you've come across on your own on Isaac Physics, you know that you're missing something somewhere.

That's where this book comes in. It isn't designed to teach you what we mean by momentum, or what a wave is. Instead, it aims to help you get a deeper understanding of the physics you've learnt by actively using that knowledge to solve problems. I hope that by doing this you'll gain insight as to why all of the ideas that you've learnt about are useful, and start to see physics as more of a unified whole.

After short chapters on general problem solving and using the hints on Isaac Physics, the majority of the book will focus on worked examples of problems which loosely follow the Isaac Physics style (particularly those problems in levels 2 to 5) but have each been written to highlight a particular concept or method that can be useful for similar parts of other problems.

You may find it useful to work through the problems inside this book while reading through it. We recommend using a pen/pencil and paper—technology is excellent, but writing and drawing on paper may help you to remember more of what you've done. Good luck.

1.1 General Problem Solving

When you first come across a new physics problem, the temptation can be to get stuck straight in with the equations. A mass here, a velocity there, and you're away calculating kinetic energy. Or should you be working out the momentum? Or possibly you are looking for an acceleration, and the only reason you've got a velocity is because one of the forces relied on the velocity. It's time to take a step back.

First of all, **read the question**. All of it. I wish I didn't have to say that, but I've come across too many students missing out on crucial marks in an exam due to having misread a question or having not actually finished reading it at all. As you go along, it can be useful to consider what some of the words mean. Words like "light' or "ideal" have particular meanings in these physics problems—you can see more of these at the back of the book. You should also be considering what physical concepts you might need to use to solve the problem—reading the entire question rather than just a small part of it can help here, as the way the question is written

1

can point towards particular ways of solving the problem.

The next step is to extract all of the **useful information** from the question. What is the question asking for? What information does it give you? Students are often told to underline or highlight the key words in a question. This is a good method, provided you know what the key words are. As well as these words, are there any physical quantities given in the question? These could be given as a numerical value with a unit, symbolically, as a description in text (look out for words like "stationary" or "at rest") or as part of a relationship between two quantities ("the pulling force is twice the magnitude of the weight of the block").

To pull all this information together, it's generally a good idea to draw at least one **diagram**. While there are a few questions where the situation is simple enough not to need a diagram or (even more rarely) too complicated or too abstract for a diagram to be sensible, these are far outnumbered by questions where a diagram would be useful. The diagram doesn't have to be a work of art, just understandable. This means that while it's totally acceptable to draw a car or a lorry as a blob, it should be a labelled blob. Drawing diagrams bigger than they need to be is better than drawing them too small and making them hard to read. Where you use letters to represent quantities, it's generally a good idea to use an "obvious" letter (like m for mass), but it's more important to make sure that you can't confuse different things. This means that you should follow these guidelines:

- Use different symbols (which can be letters with subscripts, such as m_e for the mass of an electron) for each quantity that may be involved;

- Avoid writing numerical values directly on the diagram—add a legend next to it if needs be;

- Make sure to write down next to the diagram what each letter means if the diagram doesn't make it clear what all the letters refer to;

- Make sure that letters are clearly written and placed on the diagram so you can't confuse what the letter is or which part of the diagram it refers to.

If you find that you can't fit all of the useful information you have on one diagram, then draw multiple diagrams. You may want to draw diagrams showing specific objects or sets of objects or looking at the same object from different angles. Sometimes you might draw forces on one diagram and distances on another. Whatever you decide, you should be thinking about making sure that your diagram is as clear as it can be. It is there to help you with solving the problem, as well as helping anyone else who is going to read your solution to understand your work.

Now you're nearly ready to start manipulating equations. While by this point you probably won't be able to look at the problem and work out what needs to be done to solve it, you should hopefully be able to identify some useful things that you can try. With this loose "plan" in mind, gather the equations that you think might be useful and start working through the problem. Don't be afraid to go back and try something else if you are struggling. One of the key skills of "problem solving" is recognising when something you are trying is not going to work and moving to another method instead. Make sure that you don't give up at the first sign of trouble though. The problems on Isaac Physics are designed to stretch you, and so you are likely to spend a fair amount of time feeling uncertain about what to do next, or even frustrated. Feel free to go back to earlier sections in this guide too. Is there a piece of information that you have now found that you need which may be given by the text of the question? Has your working demonstrated that you've drawn something incorrectly on your diagram and you need to redraw it? Is there another equation which you can use to take some quantities you know and turn them into a quantity that you're looking for? Might you even need several steps using different equations to do that?

These steps are often easier said than done, but with practice you will hopefully be solving problems with a lot more confidence.

1.2   Using the Hints on Isaac Physics

Sometimes people find the hints on Isaac Physics a little weird. They ask why some of the hints are just the information from the question again and then that same information added to a diagram. What they are missing is that most of the hints are designed not just to help you answer the problem that you are on, but also to get into good habits for solving more problems in the future. We hope that as you practise answering problem-solving questions you will come to use the hints less and less.

The hints for most of the questions follow the same structure, with 5 distinct types of hint:

1. Glossary and Concepts

2. Useful Information, Information Calculated Previously and Information Assumed

3. Diagram

4. Useful Equations

5. Hint Video

Some questions may omit one or more of these hint types if they aren't particularly relevant to that question. In Level 6 questions, which go beyond the A-level specifications, we also often do not use this hint structure, instead preferring to give a short hint to how to start to solve the problem or a non-obvious "trick" that you may need either to solve the problem or to make it easier. This guide will focus on the 5 standard hints.

Hint 1 - Glossary and Concepts

This hint is here to help you get started thinking about what branches of physics you may need to solve the problem. You may also find this useful to check that you've covered the appropriate concepts in your studies so far to avoid the frustration of working on a problem before you are well equipped to deal with it.

The words in the glossary are terms which have particular meanings in physics. Some of them, such as "centripetal", are terms which you may not have come across in everyday life but are regularly used in physics to describe a particular concept. Other words in this section describe the assumptions and approximations that are made in a particular problem in order to make it possible to solve without unreasonable levels of calculation. For example, surfaces are often described as "frictionless", even though in the real world there would be a small amount of friction between that surface and another surface moving relative to it. In a physics problem this word means that the magnitude of any frictional force that arises is sufficiently small that by ignoring it we can get a good approximation to the real situation which may be refined further in the future.

All of the concepts given link to a Concept Page on Isaac Physics. These pages are written as helpful reminders for people who have previously studied these concepts, not as a way of learning about a concept for the first time. The top of the page will generally contain a brief description of that concept along with one or more related equations. The sections lower down the page will elaborate on particular aspects of that concept, and the levels given on each section give an idea as to which level of question they may be required for. In some cases the links from the hint go directly to a section of the Concept Page that will be particularly useful for that question, but you may also need to read other sections on that Concept Page.

Hint 2 - Useful Information, Information Calculated Previously and Information Assumed

The second hint is the one that confuses the most people. Why is it just re-stating information given in the question? The answer is that a common problem we've seen when students get stuck with a question is that they haven't managed to extract a key piece of information from the question. This happens for a number of different reasons:

- They've failed to realise that a particular bit of text is giving them that information (saying something is massless, for example);

- They haven't realised that although a value isn't given for a piece of information, it can still be used as a symbol that may cancel out;

- They've just missed it when reading through the question.

Use this hint to check that you've got everything you need to tackle the problem.

Hint 3 - Diagram

As described in the last chapter, drawing a diagram is a key step towards solving most problems. This hint is designed to offer "faded scaffolding": at lower levels it will be a full diagram, ready to use, while at higher levels the diagram may not be drawn in the best way or may be missing some key information. Sometimes at higher levels a diagram is not given—this is because by this point you should be in the habit of drawing one yourself.

Even if there's already a diagram here, it's a good idea to draw and label your own so that you are confident about what all of the terms mean and that you have absorbed all of the information given in the question.

Hint 4 - Useful Equations

This section is pretty self-explanatory—these are some equations which you may find useful. There are often multiple ways to solve a physics problem, so you may not actually need to use all of these equations. It's also important to be clear about what the terms mean. For example, are they talking about a particular force or a resultant force? When a term such as mass appears in an equation, are you using the mass of the correct object? You may also want to think about using the same equation multiple times—does the situation in the problem change over time? Do you need to apply the equation to several different objects or collections of objects? Again, make sure that you substitute correctly into the equation.

Hint 5 - Video

The videos generally show some steps involved in solving the problem with the remainder left for you to complete. In lower levels, particularly Level 1, this may be a nearly complete solution with only the last substitutions not completed. As the levels increase, a smaller part of the solution is shown. It's a good idea to work through the steps shown in the videos yourself rather than just copying down the final state of the video. This will help you to get a better idea of what steps might be useful for similar problems in the future.

Many problems can be solved in multiple ways. Some hint videos only show one method to solve the problem while others will show a couple of methods. Don't be afraid to get started using the method in the video before switching to a different method.

Feedback hints

You may not have noticed, but when you get a question wrong, there will be some feedback underneath the answer box. While this is often just "Please try again", there is more specific feedback for common wrong answers where we've managed to work out what you've done wrong. It is always worth reading this feedback, particularly because most of the common wrong answers are the result of a mistake or misconception that's common to a few different problems.

Sig figs and rounding

Another common problem that students have is with using the correct number of significant figures. The same rule applies across the site, and this is that the correct number of sig figs to use is the lowest number of sig figs in the data that you have used to calculate your answer. You can practise this at https://isaacphysics. org/gameboards#sig_fig_prac_mastery . It's also worth noting that if you use an answer from a previous question to calculate your answer to the new question, you should use the unrounded value of the old answer to avoid accumulating rounding errors. Still give your answer to the same number of significant figures at the end, but don't round until then.

How to get extra help

If you're still struggling with some particular questions, you can get in touch with us via Twitter (@isaacphysics) or by using the contact form on our website: https://isaacphysics.org/contact. We will generally try to help you understand a particular problem or two, but we won't teach you a topic or provide answers to the problems. This book came about partly due to people requesting worked solutions to problems to help them understand how to answer particular types of question.

Mechanics

Question

A jenga tower is built consisting of 7 identical blocks, each of mass m, arranged in 4 levels. The bottom two levels contain one block each, the next level up contains two blocks with a one-block gap between them, and the top level consists of 3 blocks stuck together. The tower rests on a table and is stationary. Calculate the magnitude of these forces:

    i   the normal force acting upwards on the set of top blocks from just one of the blocks on the layer below.

    ii the normal force acting upwards on one of the two blocks on the second layer down from the block on the layer below.

    iii the normal force acting upwards on the single block on the third layer down from the block on the bottom layer.

    iv the normal force acting upwards on the single block at the bottom of the tower from the table.

Main skills required

This question is an example of a statics problem. This means that, in order to solve it, you are going to need to be confident with drawing free-body force diagrams (diagrams in which only one "body" is drawn and only the forces acting on that body are shown) along with labelled forces acting on the body. You will also need to be able to apply Newton's First Law of Motion, as this describes the forces acting on a body which is stationary, and Newton's Third Law of Motion in order to link the magnitudes of the forces on touching blocks. However as this situation is one-dimensional you only have to deal with forces acting vertically, so you don't need a full understanding of vectors.

Worked Solution

The first thing to do is to draw the whole situation out with the masses of various parts labelled, as shown in Figure 2.1. I went back and labelled the layers with letters

after realising it could be hard to refer to a specific layer otherwise. Don't be afraid
to add more details to your diagrams later.

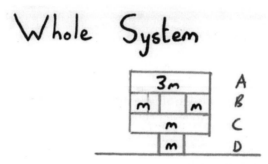

Figure 2.1. The entire Jenga tower, with the top three blocks treated
as a single object. The masses of each element are shown, along with
labels for each layer

This problem can be solved in a few different ways. One way is to look at the
forces on each layer of blocks while another method involves splitting the tower
into different sections and only looking at "external" forces acting on that section.
I'm going to start with the former and then show how the latter gives the exact same
answers (as we should expect!).

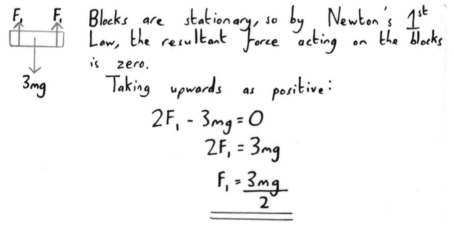

Figure 2.2 Looking at just the top layer of blocks.

Looking at just the top layer of blocks (A) in Figure 2.2, we need to show all of the
forces acting on this layer. We don't need to show any forces between the blocks
on the same layer, as when we're considering the entire layer these are "internal"

forces and would sum to zero across the entire layer[1]. We do need to include the force due to gravity, of magnitude $3mg$, acting on this layer (A) of blocks and two normal reaction forces from the two blocks below. I've labelled both of these normal reaction forces as F_1 in Figure 2.2 because they must be the same on the two sides, as the situation is symmetric. By saying that "the situation is symmetric", I mean that there's nothing that you can point to in this situation that would allow you to distinguish between the two blocks below the top layer, so they must each exert the same magnitude force on the layer above.

Another thing to notice here is that I've labelled all of the forces (which are vectors) with just their magnitude, leaving it to the direction of the arrow that I've drawn to define their direction. This means that we have to be careful to take into account their directions during the calculations, and if we ever obtain what looks like a negative magnitude, it means that the force is actually acting in the opposite direction to the arrow. In this solution I've chosen to define upwards as positive and don't have to worry about the horizontal direction as there are no components of any of these forces acting horizontally. This allows us to use Newton's 1st Law to solve for the magnitude of the force F_1.

Now we can move on to looking at the second layer (B) in Figure 2.3. Rather than looking at the whole layer, I've chosen to look at just one block, as the situation will be exactly the same for the other block due to the symmetry of the system[2]. It's very important to make sure that all of the forces acting on this block are included. As well as the weight of the block and the normal reaction force from the layer below (which I've labelled as F_2 to make it obvious that it's a force[3]), there's also a normal reaction force from the layer above. This has to exist because of Newton's 3rd Law— as each block on layer B exerts an upwards force on the layer above, the layer above must exert a force downwards (*i.e.* in the opposite direction) on each of the blocks in layer B. These downward reaction forces on layer B must each be of the same magntiude as one of the upward reaction forces on layer A, which is F_1.

All of these forces actually act in a line, but it's important to be able to distinguish the forces in the diagram, so they are all drawn slightly offset. If the question involved rotation, the correct positions would need to be shown more clearly on a larger diagram.

When using Newton's 1st Law to calculate F_2, we had to substitute in for the value of F_1 from the previous part of the question. The reason I didn't immediately use $3mg/2$ here but initially wrote F_1 is so that it's clear in my working where the $3mg/2$ came from. Although it's pretty obvious in this situation, it's a good habit

---

[1] Otherwise the layer would be accelerating sideways, as can be seen from Newton II

[2] Remember that this means that there's nothing in the system which would enable me to distinguish between these two blocks on this level—they are equivalent to each other.

[3] While you can label using whatever letter you want, it's generally a good idea to choose a letter to make it obvious what it is that you're labelling, whether that's a force, a mass, a velocity, *etc.*

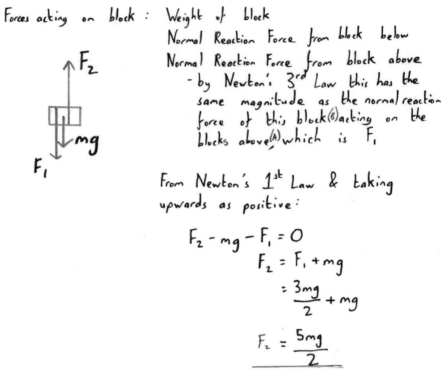

Forces acting on block : Weight of block
Normal Reaction Force from block below
Normal Reaction Force from block above
 - by Newton's 3rd Law this has the
 same magnitude as the normal reaction
 force of this block (B) acting on the
 blocks above (A) which is F$_1$

From Newton's 1st Law & taking
upwards as positive :

$$F_2 - mg - F_1 = 0$$
$$F_2 = F_1 + mg$$
$$= \frac{3mg}{2} + mg$$
$$F_2 = \frac{5mg}{2}$$

Figure 2.3. Looking at one of the two blocks on the layer below the top row (Looking at row B)

to get into—make sure that you can follow your own working when you look back before you move onto more complicated situations.

The rest of this problem can be solved in a similar way, remembering always to include the normal reaction forces from both above and below the current layer - for layer C as shown in Figure 2.4 these normal reaction forces are from other layers of blocks, but for layer D as shown in Figure 2.5, the normal reaction force acting upwards is from the surface the tower is resting upon.

As I said earlier, it's also possible to solve this problem by selecting just some parts of the set of things we're looking at, which we call a "subsystem"—in this case our subsystem is a layer of the tower. By using this method, we can make it so that our subsystem always extends to the top of all the blocks, so we never have to consider a normal reaction force acting downwards, as there is nothing sitting on top of our subsystem of blocks. All of the reaction forces here are labelled the same as in the previous method. We can start with the simplest "subsystem" - the entire tower, as shown in Figure 2.6.

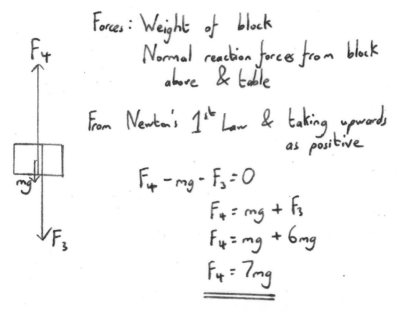

Forces : Weight of block
         Normal reaction forces from
             block below & blocks above

From Newton's 1st Law, upwards as +ve

$$F_3 - mg - 2F_2 = 0$$
$$F_3 = 2F_2 + mg$$
$$= 2\left(\frac{5mg}{2}\right) + mg$$
$$\underline{\underline{F_3 = 6mg}}$$

Figure 2.4. Continuing to solve the rest of the problem by looking at layer C

Forces : Weight of block
         Normal reaction forces from block
             above & table

From Newton's 1st Law & taking upwards
                                as positive

$$F_4 - mg - F_3 = 0$$
$$F_4 = mg + F_3$$
$$F_4 = mg + 6mg$$
$$\underline{\underline{F_4 = 7mg}}$$

Figure 2.5. Continuing to solve the rest of the problem by looking at layer D

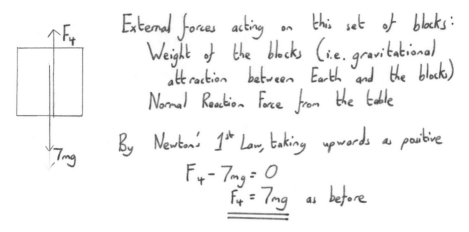

Figure 2.6 Considering the entire tower as a single system

All of the weight of the blocks included in the subsystem can be thought of as acting through the centre of gravity of the subsystem. We can ignore the normal reaction forces between blocks inside the subsystem because we are treating the subsystem as if it were one object, so the equal and opposite normal reaction forces cancel each other out[4]. It's important to realise that these forces act on different blocks, so normally we'd have to treat them separately and they wouldn't cancel out[5], but since here we're looking at a collection of blocks, these "internal" forces cancel out on the subsystem.

We can then take a "subsystem" consisting of the top three layers of blocks (A, B and C), as shown in Figure 2.7, and a "subsystem" consisting of just the top two layers of blocks (A and B), as shown in Figure 2.8. To finish off the problem we'd then have to take just the top layer of blocks, and this is exactly the same as we've already done in Figure 2.2.

It's up to you which of these two methods you are most comfortable with, but it's worth becoming familiar with both, as some situations are easier to solve with a particular method.

---

[4]We can use the same argument as we did before with the blocks glued together along with Newton's Third Law to show that these must cancel out.

[5]A common mistake with Newton's Third Law is to think that the "equal and opposite" forces act on the same object and so give zero resultant force. These action-reaction pairs have to act on different objects. If you are balancing out forces on a single object, you are probably using Newton's 1st Law.

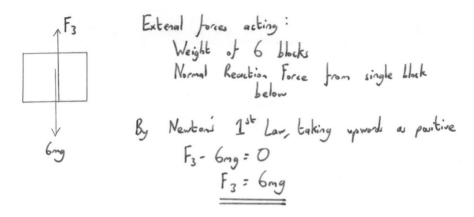

F_3

$6mg$

External forces acting :
    Weight of 6 blocks
    Normal Reaction Force from single block
        below

By Newton's 1st Law, taking upwards as positive

$$F_3 - 6mg = 0$$
$$F_3 = 6mg$$

Figure 2.7 Considering the top three layers as a single subsystem

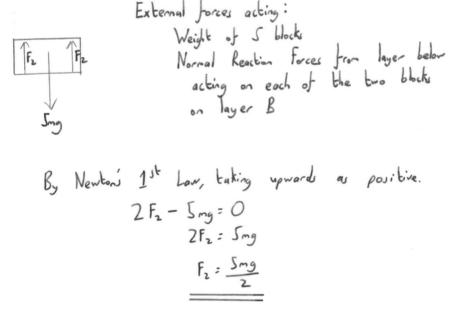

F_2 F_2

$5mg$

External forces acting :
    Weight of 5 blocks
    Normal Reaction forces from layer below
    acting on each of the two blocks
    on layer B

By Newton's 1st Law, taking upwards as positive.

$$2F_2 - 5mg = 0$$
$$2F_2 = 5mg$$
$$F_2 = \frac{5mg}{2}$$

Figure 2.8 Considering the top two layers as a single subsystem

Extra things to think about

All of the previous working assumes that the situation is symmetrical. How would the normal reaction forces change if one of the blocks on layer B has a larger mass than the other, but the sizes and shape of the blocks remain the same?

2.2   2D Statics - Box on a Rough Surface

Question

A box of mass 275 g sits on top of a flat, rough surface. A child attempts to pull it with a horizontal force of 1.2 N, but the box remains stationary. What are the magnitudes of:

  i The normal reaction force from the ground acting on the box?

  ii The frictional force acting on the box?

Main skills required

This question demonstrates the independence of forces acting in perpendicular directions. It uses Newton's 1st Law and the ability to convert between g and kg. A knowledge of what is meant by significant figures is also needed in order to explain why the final answers are as stated.

Worked Solution

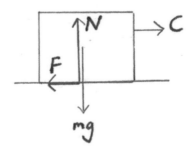

Figure 2.9 The forces acting on the block

All of the given information should first be placed into a diagram to make it all easily accessible. In Figure 2.9 all of this has been represented purely by symbols, rather than the values given. This is to avoid clutter on the diagram while still making it clear what each of these symbols is referring to. The letters used have been chosen to attempt to make it obvious what each letter represents, so F represents a friction force, N the normal reaction force, and C the force from the child. When choosing symbols, you should think about how best to represent all of the pieces of information in the question, which may include forces, velocities, displacements, masses, charges and so on[6]. It's important to be careful about the direction of the frictional force—it acts in a direction so as to oppose the relative motion of the two surfaces[7]. As the child is pulling to the right in my diagram, in the absence of friction this would cause the block to accelerate to the right relative to the ground, so the frictional force acts to the left to oppose this.

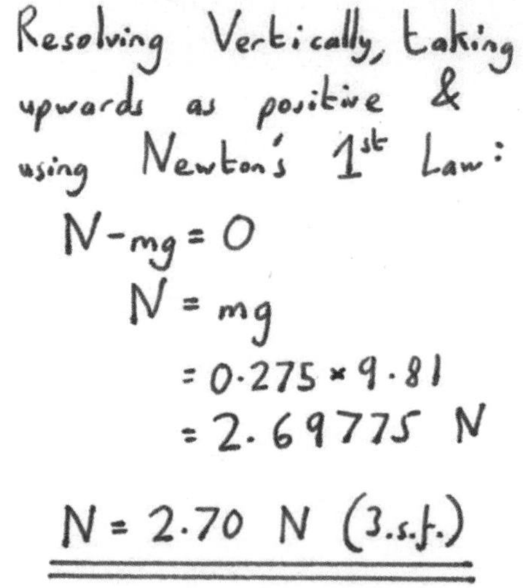

Resolving Vertically, taking upwards as positive & using Newton's 1st Law:

$$N - mg = 0$$
$$N = mg$$
$$= 0.275 \times 9.81$$
$$= 2.69775 \text{ N}$$

$$\underline{N = 2.70 \text{ N} \quad (3.s.f.)}$$

Figure 2.10 Resolving forces vertically

A good step after this is to think about what physical laws we can use. As the box remains stationary, it is not accelerating, and therefore we can use Newton's First Law to show that there must be no resultant force acting on the box. Rather than try to deal with all of the forces at once, we can instead look at two perpendicular directions; the directions must be perpendicular to ensure that a force in one of those directions has no component acting in the other direction, allowing us to treat them independently. For each of these directions, in this case vertically and horizontally[8], a decision must be made as to in which direction to take the forces as

---

[6]Generally speaking, u is used to represent a velocity before some change, and v used to represent the final velocity. You should use subscripts to represent different bodies, so u_1 would be the velocity of object 1 "before", and v_2 would be the velocity of object 2 "after".

[7]It's particularly important to realise that this is not necessarily against the motion of the overall object. When a car accelerates, the frictional force acting on the wheels from the road is in the direction that the car accelerates—this is how a car can move at all!

[8]While we could have chosen any two perpendicular directions, the fact that the forces acting in this question are all either horizontal or vertical means that this should be the ob-

positive[9].

In Figure 2.10, I've applied Newton's First Law to the vertical components of the forces acting on the box, which in this case is just two vertical forces: the weight of the box and the normal reaction force from the floor. It's important to note that as the mass of the box was given in g, it should first be converted into kg, as the SI base unit of mass is the kilogram (this is the only SI unit where the base unit contains a prefix). If you don't convert the mass, your answer would not be in newtons[10] (as $1\,N = 1\,kg\,m\,s^{-2}$), and an answer of 2700 N is incorrect.

I can similarly find the frictional force by using Newton's First Law and considering only the horizontal components of the forces, as shown in Figure 2.11.

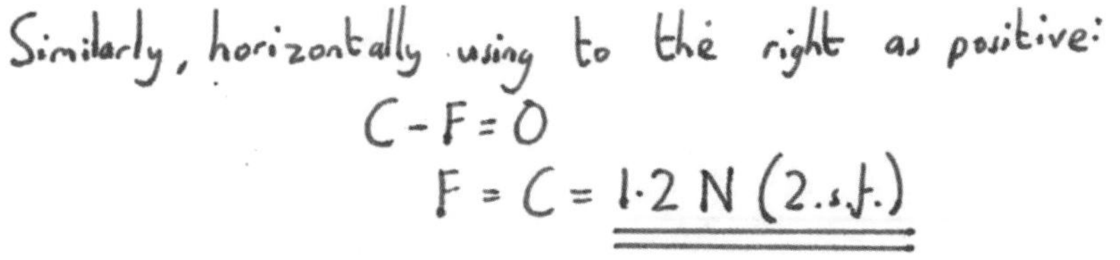

$$\text{Similarly, horizontally using to the right as positive:}$$
$$C - F = 0$$
$$F = C = 1\cdot2\ N\ (2.s.f.)$$

Figure 2.11 Resolving forces horizontally

You'll notice that the answer for the normal reaction force was given here to 3 significant figures, as the mass of the box is given to 3 significant figures. The magnitude of the pulling force that the child applies to the box is only given to 2 significant figures, but as this force does not affect the magnitude of the normal reaction force, we don't have to worry about it; however, when we calculate the frictional force, we do use the child's pull, so we can only answer to 2 significant figures.

Extra things to think about

The child stands up, so their pull now acts at some angle to the horizontal. How would the direction of the pulling force affect the normal reaction force from the ground acting on the box and the frictional force needed to stop it from moving?

---

vious choice for this problem.

[9]A force in the other direction would then be considered as negative.

[10]While it would be technically correct to give answers to this question in units of $g\,m\,s^{-2}$, physicists would generally try to avoid using this unit. They prefer to use more familiar units made from SI base units.

2.3   2D Statics - Varying Friction

Question

> A bag of grain of mass $m = 100$ kg (to 3.s.f.) needs to be moved across a ware-
> house, but the person pulling it can only apply a horizontal force $P = 450$ N
> (to 3.s.f.) to move it. A frictional force with a static coefficient of friction of $\mu = 0.55$ opposes this motion. A rope is attached to the top of the bag of grain
> through a frictionless pulley system so that an upwards force can be applied
> to the bag. What is the minimum magnitude of this force which would allow
> the bag to be moved sideways?

Main skills required

This question also uses the independence of forces acting in perpendicular directions. It uses Newton's First Law, as well as the equation for the maximum possible
static friction acting between two surfaces, $F_{max} = \mu N$. It also requires you to think
about a "limiting case"—the point at which the behaviour of the system changes
sharply.

Worked Solution

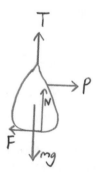

Figure 2.12 The forces acting on the bag

There are actually five forces that you need to consider for this question, so it's
important to think carefully about all of these and draw a diagram like that in Figure 2.12 to represent all of these forces acting on the bag. Three of the forces are
explicitly described in the question—the upwards force from the rope which we're

trying to find (labelled on the diagram as T as this force is a tension[11]), a frictional force (labelled on the diagram as F) and the pulling force P. In addition to these, we also need the weight of the bag, as we assume that this whole situation is taking place on Earth, and the normal reaction force from the floor acting on the bag since they are touching[12].

The next step is to consider what specific situation we care about – it's obvious that if a large enough upwards force were applied to the bag, it would be possible to push it sideways. This upwards force would cause the bag to come off the floor, meaning there was no friction at all between the bag and the floor; however, the question asks for the minimum, so we should be looking at a situation where the pulling force is just big enough to overcome the friction: the "limiting case" where the bag is on the boundary between remaining stationary and moving. At this point:

- The resultant force on the bag is zero (if it was not zero the bag would already be accelerating)
- The frictional force is at its maximum – if the frictional force was not at its maximum, it could increase to counter an increased pulling force.
- Any increase in the upwards force would allow the bag to move sideways, as this would reduce the magnitude of the normal reaction force, causing the maximum frictional force to decrease to lower than the applied pulling force.

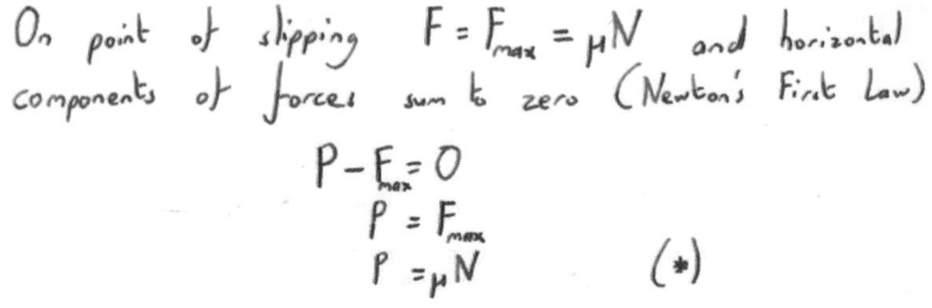

On point of slipping $\quad F = F_{max} = \mu N$ and horizontal components of forces sum to zero (Newton's First Law)

$$P - F_{max} = 0$$
$$P = F_{max}$$
$$P = \mu N \qquad (*)$$

Figure 2.13 The horizontal components of forces acting on the bag.

By applying Newton's First Law at this point (to explain why the resultant force must be zero), as shown in Figure 2.13, we can equate the magnitudes of the horizontal forces.

Similarly, as shown in Figure 2.14, we can apply Newton's First Law to the vertical forces in order to work out the normal reaction force, which we need in order to be able to find the magnitude of the frictional force.

---

[11] As usual, you could use whatever letter made most sense to you, avoiding those given in the question and any others which may lead to confusion.

[12] I've deliberately drawn the normal reaction force slightly to the side of the weight here. Can you work out why? You may need to think about moments.

Resolving vertically, using Newton's First Law, taking upwards as positive:

$$T + N - mg = 0$$
$$N = mg - T$$

Figure 2.14 The vertical components of forces acting on the bag.

Substituting this N into the equation $(*)$

$$P = \mu(mg - T)$$
$$\frac{P}{\mu} = mg - T$$
$$T + \frac{P}{\mu} = mg$$
$$T = mg - \frac{P}{\mu}$$
$$= (100 \times 9.81) - \left(\frac{450}{0.55}\right)$$
$$= 162.8 \, N$$
$$\Rightarrow 160 \, N \ (2 \text{ s.f.})$$

Figure 2.15 Combining our equations to solve the problem.

Now that we've fully described the relevant physics of this situation, we can combine our equations to get an equation for the magnitude of the upwards force T. Substituting in the values from the question (Figure 2.15) allows us to give a value for this force to only 2 significant figures, as although most of the values are given to 3 significant figures, the coefficient of friction is given to a lower number of significant figures and therefore places a limit on the precision of our answer.

Extra things to think about

How would you solve this problem if, instead of changing the upwards force from the rope, the person pulling the bag could pull at a range of different angles? When the bag starts to move, the friction changes from static friction to dynamic friction which generally has a lower coefficient of friction. What does this say about how the force from the rope could be changed while still allowing the bag to move?

2.4   1D Kinematics - Two Balls Falling Simultaneously

Question

A ball is thrown vertically upwards so that it reaches a height h_0 above the ground. Another ball is placed on a shelf at half this height. The second ball is then knocked horizontally off the shelf and hits the ground at the same time as the first ball. Ignoring air resistance, at what height is the first ball when the second ball starts to fall?

Main skills required

This question requires knowledge of the SUVAT equations of motion, or how to calculate velocity and displacement for constant acceleration. Defining an "initial condition" for the problem is made easier if conservation of energy is used. This question also uses the ability to recognise and solve simultaneous equations as well as the quadratic equation formula.

Worked Solution

The two balls do not interact, so the diagram showing them both can be thought of as two different diagrams. As usual, we put the key info on these diagrams, shown in Figure 2.16. Both situations rely on the value of h_0, so it's important to make sure we don't change this definition between the two diagrams.

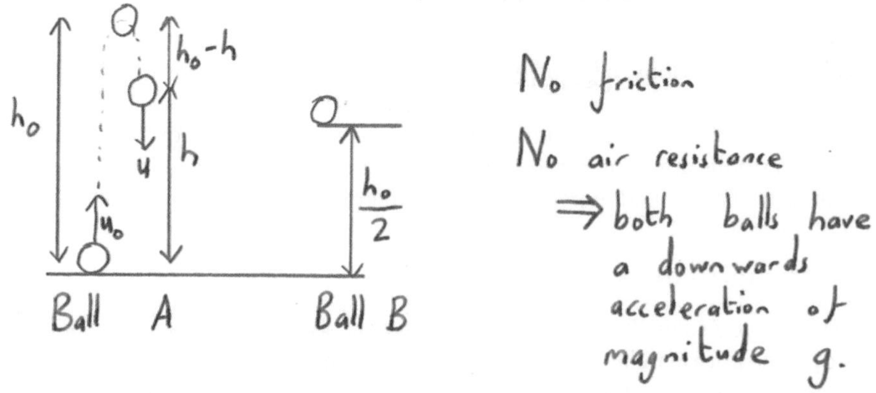

Figure 2.16. The movement of both balls up until the point where ball B is knocked off the shelf.

If the two balls hit the ground together, then the time for ball A to fall from some height[13] h, with an initial speed downwards of u, **will be the same** as the time for ball B to fall from the shelf to the ground, starting from rest (we can ignore the horizontal motion of ball B as it won't affect our answer at all!). To find the time that ball A takes to fall from this height, we have to calculate u[14]. Since we know the maximum height it reached, h_0, at which point its velocity must have been zero (as we're at a maximum and know it's not moving horizontally), we can look solely at the motion after this point until it reaches the height h. This can be done either by looking at the acceleration of the ball and using the equations of motion, or by conserving energy[15] as in Figure 2.17.

Using conservation of energy to find the speed of the ball at height h :

$$KE + GPE \text{ at height } h = GPE \text{ at top (height } h_0)$$

$$\frac{1}{2}mu^2 + mgh = mgh_0$$

$$\frac{u^2}{2} + gh = gh_0$$

$$\frac{u^2}{2} = g(h_0 - h)$$

$$u^2 = 2g(h_0 - h)$$

$$u = \sqrt{2g(h_0 - h)}$$

Figure 2.17 Using conservation of energy to find u.

We now have a starting point which can be used in our SUVAT equations. To make sure I was being clear both in my mind and my explanation, I've drawn a new diagram, Figure 2.18 to represent just the motion of the ball from height h downwards. You may note that I've defined downwards as positive now, as the motion of the ball is entirely downwards[16].

---

[13] This is the height we are asked for in the question.

[14] Actually we could instead calculate v and use a different SUVAT equation, but here I've chosen to find u rather than v.

[15] It's quite common in physics problems to be able to use some different methods to get the same answer. You can use a couple of different methods to find the same answer to check

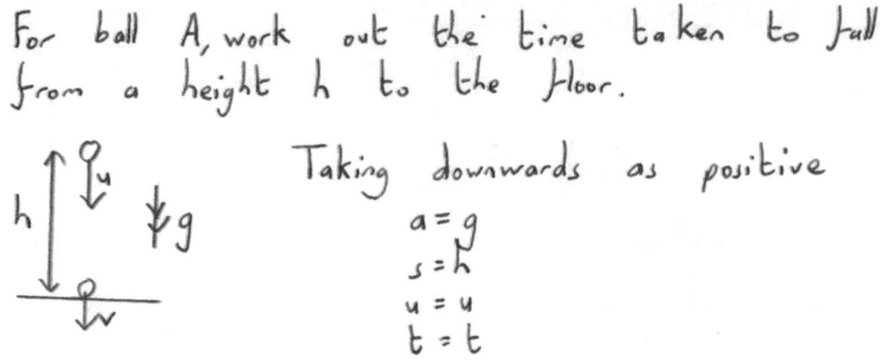

For ball A, work out the time taken to fall
from a height h to the floor.

Taking downwards as positive

$$a = g$$
$$s = h$$
$$u = u$$
$$t = t$$

Figure 2.18 Finding the time for ball A to fall from a height h.

In Figure 2.18 and while working through using the SUVAT equations (Figure 2.19) I've chosen to keep the term u rather than subsitute in for u in terms of g, h_0 and h. This will just help us keep our working tidier, but it would be just as correct to substitute in for u here.

$$s = ut + \tfrac{1}{2}at^2$$
$$h = ut + \tfrac{g}{2}t^2$$
$$\frac{2h}{g} = \frac{2u}{g}t + t^2$$
$$t^2 + \frac{2u}{g}t - \frac{2h}{g} = 0 \qquad (*)$$

Figure 2.19 Rearranging the SUVAT equation for ball A.

your working!

[16]Of course, you could just as easily define upwards to be positive, provided you re- membered that the direction of ball A's displacement from this point, velocity and accel- eration are all downwards, so these would then be represented as negative numbers.

For ball B we can again use the SUVAT equations of motion but with an initial velocity of zero as the ball was not moving vertically at the beginning of its fall[17]. This is shown in Figure 2.20, giving us another equation involving t, the time for which both balls were falling.

In the same time t, ball B must fall a distance $\frac{h_0}{2}$, having started from rest.

$$\downarrow = +ve$$

$$a = g$$

$$s = \frac{h_0}{2}$$

$$u = 0$$

$$s = ut + \tfrac{1}{2}at^2$$

$$\frac{h_0}{2} = 0 + \frac{gt^2}{2}$$

$$\frac{h_0}{g} = t^2$$

$$t = \sqrt{\frac{h_0}{g}}$$

Figure 2.20 Using the equation of motion of ball B to find t.

This value for t can be substituted back into the equation marked with a star earlier, as shown in Figure 2.21, because t refers to the same time for each of the two situations[18].

---

[17]I'm again defining downwards to be positive, but as before you can choose to keep upwards as positive.

[18]Being able to translate from the words in the question to statements like this about the two times being the same is a key skill for solving these problems.

This value for t can be substituted into the equation marked (*) earlier.

$$t^2 + \frac{2u}{g}t - \frac{2h}{g} = 0$$

$$\frac{h_0}{g} + \frac{2u}{g}\sqrt{\frac{h_0}{g}} - \frac{2h}{g} = 0$$

$$h_0 + 2u\sqrt{\frac{h_0}{g}} = 2h$$

Substituting in the value of u obtained earlier:

$$h_0 + 2\sqrt{2g(h_0-h)}\sqrt{\frac{h_0}{g}} = 2h$$

$$h_0 + 2\sqrt{2h_0^2 - 2h_0 h} = 2h$$

$$\sqrt{2h_0^2 - 2h_0 h} = h - \frac{h_0}{2}$$

Figure 2.21 Substituting back in for t and u.

The rest of the problem is just algebraic manipulation to find the value of h in terms of h_0. As shown in Figure 2.22, this involves both squaring both sides in order to remove a square root and using the quadratic formula. It's often quite easy to get lost in manipulations like this, so don't be afraid of going back to an earlier part of your working if it seems that you are going around in circles or just aren't making progress. Your knowledge of what to do will improve with experience—in other words, attempting lots of questions.

Squaring both sides

$$2h_o^2 - 2h_o h = h^2 + \frac{h_o^2}{4} - hh_o$$

$$h_o^2\left(2-\frac{1}{4}\right) = h^2 + h_o h$$

$$h^2 + h_o h - \frac{7}{4}h_o^2 = 0$$

$$h = \frac{-h_o \pm \sqrt{h_o^2 + 7h_o^2}}{2}$$

$$= -\frac{h_o}{2} \pm \frac{\sqrt{8h_o^2}}{2} = -\frac{h_o}{2} \pm \sqrt{2}h_o$$

h cannot be negative so

$$h = h_o\left(\sqrt{2} - \frac{1}{2}\right)$$

Figure 2.22 Solving the problem by rearranging to get h.

Extra things to think about

- At what height would the shelf need to be for the two balls to hit the ground at the same time if ball B was knocked off the shelf at the point where ball A reached its maximum height?

- How would you solve this problem if ball B was not knocked off a shelf but was instead given an initial speed upwards equal to half that of ball A's initial speed? How about if B was given this same speed downards?

2.5   2D Kinematics - Trajectory of a ball

Question

> A ball is thrown from ground level at an angle θ to the horizontal. If it first hits the ground a distance x away from its initial position, what speed was it thrown at? Ignore air resistance.

Main skills required

This question requires knowledge of the SUVAT equations of motion, or how to calculate velocity and displacement for constant acceleration. It also uses the ability to recognise and solve simultaneous equations as well as being able to separate vectors out into perpendicular components.

Worked Solution

Since there is no air resistance, the ball travels along with a vertical acceleration of magnitude g towards the ground. This motion is shown in Figure 2.23.

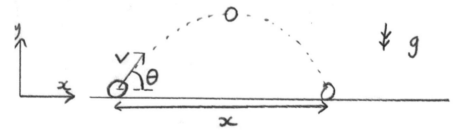

Figure 2.23 Motion of the ball after it has been thrown.

There is no acceleration horizontally, so it's useful to separate out the motion in the x and y directions[19]. It's often the case in physics problems that we have to define elements which are not given in the question, in this case the time for which the ball is in the air. This will be used in the equations for the motion in both the x and y directions. We can define this time as T, as shown in Figure 2.24.

Since the time T links the motions vertically and horizontally, we can use it to link the horizontal and vertical motion without knowing its value (as we're trying to

---

[19]Actually, for this to be the case, we only need the accelerations in the x and y directions to be independent of motion in the other direction.

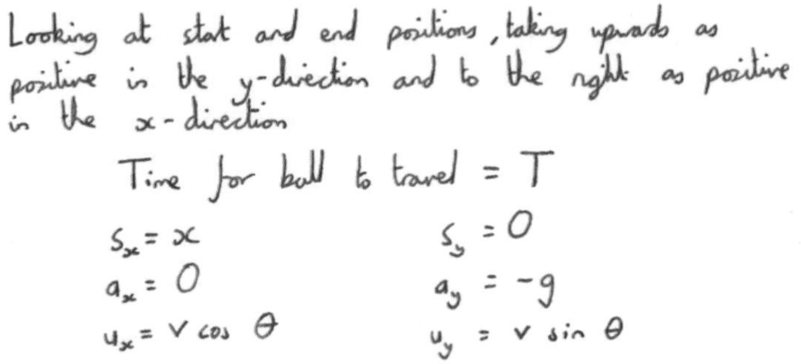

Looking at start and end positions, taking upwards as positive in the y-direction and to the right as positive in the x-direction

Time for ball to travel $= T$

$S_x = x$ $\qquad$ $S_y = 0$

$a_x = 0$ $\qquad$ $a_y = -g$

$u_x = v \cos \theta$ $\qquad$ $u_y = v \sin \theta$

Figure 2.24. Setting up the problem by separating out vertical and horizontal motion.

find the initial speed of the ball, not the time). One way of doing this is by finding an expression for T from one direction and then substituting this into an expression for motion in the other direction. For example, we can find T by looking at the vertical motion of the ball from when it is thrown to when it lands over which there is a vertical displacement of zero. This is shown in Figure 2.25. In this case the solution, $T = 0$ has to be ignored as the ball is also at this displacement when it is thrown. Also, you should notice that the u and v given in the SUVAT equations of motion do not always correspond to velocities that we've defined as u or v. It's very important when working through these problems to think carefully about which symbols you need to substitute in for the initial and final velocities.

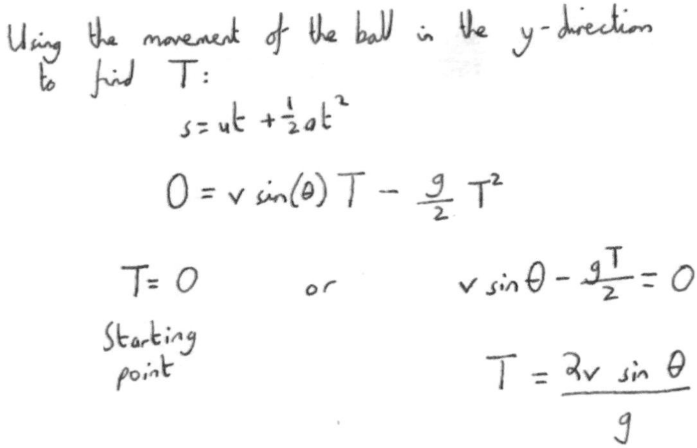

Using the movement of the ball in the y-direction to find T:

$$s = ut + \tfrac{1}{2}at^2$$

$$0 = v \sin(\theta) T - \tfrac{g}{2} T^2$$

$T = 0$ $\qquad$ or $\qquad$ $v \sin \theta - \tfrac{gT}{2} = 0$

Starting point

$$T = \frac{2v \sin \theta}{g}$$

Figure 2.25 Finding the time before the ball hits the ground.

Alternatively, we can use the symmetry of the situation—that the motion of the ball downwards is exactly the opposite of the motion of the ball upwards. This means that the ball takes a time $T/2$ to reach the top of its motion, and at this point the component of its velocity in the vertical direction must be zero. This gives us an alternative way to find T, as shown in Figure 2.26. This method gives only a single solution for T.

Alternatively, using symmetry of the situation:
At top of curve $t = \dfrac{T}{2}$

$$v_y = 0$$

$$v = u + at$$

$$0 = v \sin \theta - g\dfrac{T}{2}$$

$$T = \dfrac{2v \sin \theta}{g}$$

Figure 2.26 An alternative method of finding T, using symmetry.

In either case, this value for T can then be used in the equation of motion for the ball's horizontal motion, as shown in Figure 2.27. Rearranging this equation gives v, the initial speed of the ball, in terms of the given distance x and angle θ as well as the acceleration due to gravity g, as required in the question.

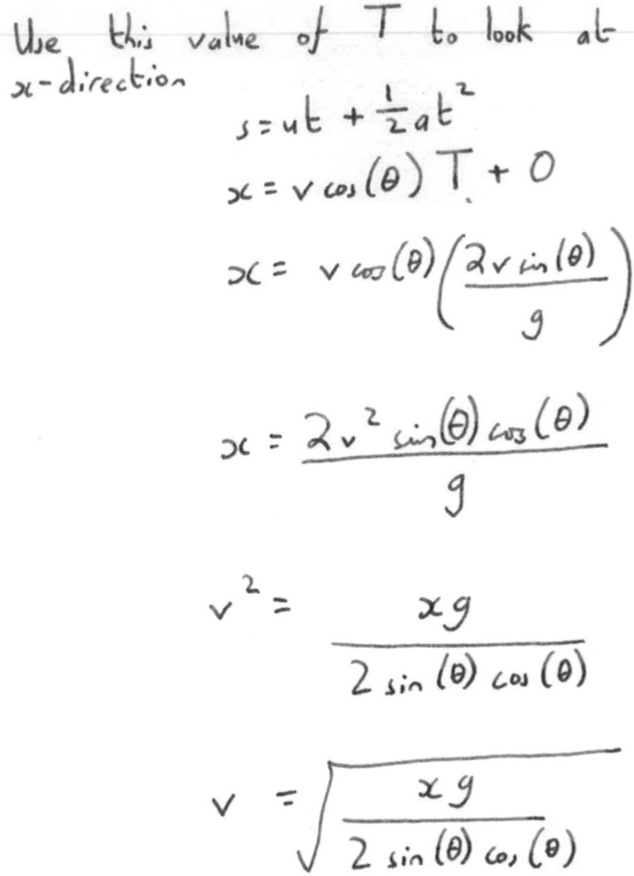

Use this value of T to look at motion in x-direction

$$s = ut + \tfrac{1}{2}at^2$$

$$x = v\cos(\theta)\, T + 0$$

$$x = v\cos(\theta)\left(\frac{2v\sin(\theta)}{g}\right)$$

$$x = \frac{2v^2 \sin(\theta)\cos(\theta)}{g}$$

$$v^2 = \frac{xg}{2\sin(\theta)\cos(\theta)}$$

$$v = \sqrt{\frac{xg}{2\sin(\theta)\cos(\theta)}}$$

Figure 2.27. Finding an equation for the horizontal distance travelled, and then using this to find the initial speed.

Extra things to think about

How would you complete the question if instead the ground sloped down, at an angle ϕ to the horizontal, away from where the ball was thrown?

2.6   2D Kinematics - Crossing a River

Question

A kayaker can travel at a speed u relative to still water. They attempt to cross a straight portion of a river in which the water is running at a speed of v_{river}, parallel to the banks.

i If they attempt to cross the river by paddling directly towards the opposite bank, what is their resultant speed relative to the riverbank that they left from?

ii If they instead adjust the direction that they paddle so that their resultant motion is directly across the river, what is their resultant speed and at what angle to the direct line across the river do they need to aim?

Main skills required

This question requires knowledge of relative velocities and how to add vectors, along with some trigonometry.

Worked Solution

We'll start with the situation where the kayaker paddles directly towards the opposite bank but is moved off course by the motion of the water. Both the kayaker's velocity relative to the water u and the water's velocity relative to the bank are vectors, which are represented in Figure 2.28 by arrows[20]. The resultant velocity of the kayak relative to the riverbank is given by the vector sum of these two; to add the two velocities, they must be drawn tip-to-tail. The magnitude of the resultant velocity can be calculated using Pythagoras' Theorem, as **the two given velocities are at right angles to each other**.

The situation is slightly different when the kayaker is trying to ensure that they move directly across the river. In that case, the vector which is at right angles to the velocity of the river is the **resultant** velocity. This means that the resultant velocity is not the hypotenuse of this right-angled triangle, and instead the vector u is the hypotenuse. As before, the magnitude of the resultant velocity can be calculated using Pythagoras' Theorem, making sure to be clear about which side is

---

[20]Real rivers have a variation in speed across their width, but physics problems often start with simplified versions of a situation so that you can use them to practise before heading on to more realistic problems.

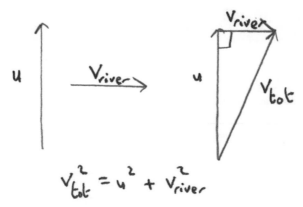

$$V_{tot}^2 = u^2 + V_{river}^2$$

Figure 2.28. Adding together the velocity of the kayak relative to the river and the river relative to the bank, to get the resultant velocity of the kayak relative to the bank.

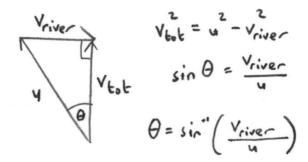

$$V_{tot}^2 = u^2 - V_{river}^2$$

$$\sin \theta = \frac{V_{river}}{u}$$

$$\theta = \sin^{-1}\left(\frac{V_{river}}{u}\right)$$

Figure 2.29. Adding together the two vectors to get a resultant that is perpendicular to the velocity of the river water relative to the bank.

the hypotenuse. The angle at which the kayaker must paddle can be found using trigonometry.

Extra things to think about

If the kayaker needed to reach a point a distance d from a point opposite their starting point along the far bank of a river of width w, in what direction should they paddle? Assume that their maximum speed and the speed of the river are the same as given previously.

How would these calculations have to change if the speed of the water in the river varied across its width?

2.7   2D Dynamics - Resolving Forces on a Slope

Question

A block of mass m sits on a frictionless slope which is at an angle of θ to the horizontal. What is the magnitude of the normal reaction force acting on the block, and what is the acceleration of the block down the slope?

Main skills required

This question involves finding the components of vectors in directions which are not horizontal and vertical in order to use them in Newton's First and Second Laws of motion.

Worked Solution

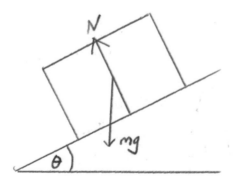

Figure 2.30 Forces acting on a block on a frictionless slope.

As usual, the first step is drawing a diagram of the situation using what we already know. One force acting on the block, the normal reaction force, is mentioned in the question—you should have expected this force to be involved due to two objects touching each other[21]. We also need to add the weight of the block, another force that you should expect. As the slope is described as frictionless, there is no frictional force[22].

---

[21] If you have studied moments, you may want to ask yourself why I've chosen this particular point on the block to draw the normal reaction force acting through.

[22] Really what we're modelling here is a surface where the frictional force would be very small, so small that it's not worth including in our calculation as it wouldn't really change the answer.

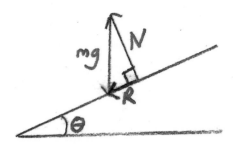

Figure 2.31. Adding the forces on the block to give a resultant force R down the slope.

In situations like this, we know that the block will not leave the surface of the slope[23]. This situation can be looked at in a few different ways. One way is to draw a diagram showing the addition of the two forces on the block to give a resultant force which must be in the direction down the slope, as shown in Figure 2.31.

As the resultant force is parallel to the slope and the normal reaction force is perpendicular to the slope, we get a right-angled triangle. We can find the magnitude of the resultant force on the block using trigonometry, provided we can find one of the angles in this triangle. By drawing a similar triangle on the diagram, like in Figure 2.32, make sure that you can show that the angle between the gravitational force on the block and the normal reaction force must be θ.

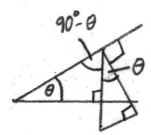

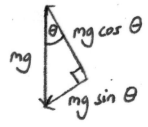

Figure 2.32. Using a similar triangle to find the angle between the block's weight and the normal reaction force acting on the block.

Figure 2.33. Using trigonometry to find the magnitudes of the normal force and the resultant force acting on the block.

---

[23]If it did start to leave the slope due to a larger normal reaction force, then as soon as it lost contact, there would be zero normal reaction force, and it would immediately fall back towards the slope again. You'd need an extra impulse to "bump" it off the slope.

Using Newton's Second Law and the
resultant force down the slope

$$F = ma$$

$$mg \sin \theta = ma$$

$$a = g \sin \theta \text{ down the slope}$$

Figure 2.34. Using Newton's Second Law to find the acceleration of the block.

You can now use the cosine of this angle to find the magnitude of the normal reaction force and the sine of this angle to find the magnitude of the resultant force acting on the block as shown in Figure 2.33. The acceleration of the block can be found using Newton's Second Law, as shown in Figure 2.34.

Another method to find these two forces involves resolving forces in two perpendicular directions. We know from the block not leaving the surface that **there is no resultant force in the direction perpendicular to the slope**, so it's useful to pick this direction to resolve forces. We know that we must be able to balance forces in this direction. We can't balance forces vertically, as the block could accelerate along the slope, and this would mean that it would have a component of acceleration vertically, so it must have a non-zero resultant force in this direction[24].

Although we know that the forces must balance perpendicular to the slope, we don't know what the force will be in the direction at right angles to this - parallel to the slope. We therefore have to look at the components of the forces in these two directions in order to solve to find the resultant force along the slope. The normal reaction force only has a component perpendicular to the slope, but we need to split the weight of the block into two components.

The relationship between the components and the angle θ is shown in Figure 2.33. Check that you can show that this angle is indeed θ by drawing suitable lines on Figure 2.30, to give Figure 2.32. A good way of checking to see if you've got some trigonometry the right way round is to think about what would happen for specific angles, like $0°$ or $90°$. In this case, an angle of $0°$ would mean that there is no component of the weight acting along the slope, as you would expect for a flat surface.

---

[24]You could solve the problem resolving forces vertically and horizontally so long as you were careful about using the correct component of acceleration in each of these directions. This is a more complicated way of solving the problem, and so more likely to lead to mistakes!

We can now use our knowledge that the block has no acceleration in the direction perpendicular to the slope, along with Newton's First Law, to obtain the magnitude of the normal reaction force, as shown in Figure 2.35.

Resolving perpendicular to slope, using Newton's First Law & defining up away from the slope as positive : $N - mg \cos \theta = 0$

$$N = mg \cos \theta$$

Figure 2.35. Looking at the components acting perpendicular to the slope.

As there's only one force with a component acting down the slope, there is an unbalanced or resultant force in this direction. As before, in Figure 2.34, we can use Newton's Second Law to convert this into an acceleration.

Extra things to think about

The block is now placed on a rough slope, so there is some friction between the block and the slope. The maximum possible frictional force is related to the magnitude of the normal reaction force by $F = \mu mg$ where μ is the coefficient of friction between the surfaces. As the angle between the slope and the horizontal increases, how would the movement (or otherwise) of the block down the slope change, and why? At what angle to the horizontal does this change occur?

2.8   2D Dynamics - Heading a Football

Question

> A football of mass m with an initial velocity $\underline{u}$ at an angle of α to the horizontal
> drops towards a footballer. The footballer then heads the ball back so that it
> has a new velocity of $\underline{v}$ at a smaller angle β above the horizontal. If the time
> that the ball is in contact with the player's head is t, what is the average force
> on the football player's head during the header?

Main skills required

This question uses vector subtraction (including some trigonometry), as well as
Newton's Second Law in its original form—given in terms of momentum. Newton's
Third Law is also needed for a proper understanding of the situation.

Worked Solution

The first thing to do is to draw the motion of the football just before and after the
header, on a diagram. The question states that the initial and final velocities are in
the same vertical plane, so we can choose to draw the situation in the plane of the
paper, as in Figure 2.36.

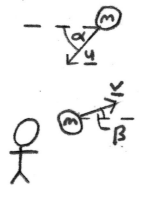

Figure 2.36 The motion of the ball before and after the header.

The next step is to recognise that the force on the ball can be linked to these
velocities through Newton's Second Law. Although the average acceleration could

be calculated and then used in the equation $\underline{F} = m\underline{a}$, it may be simpler to think about solving this problem in terms of the original formulation of Newton's Second Law so that the force is equal to the rate of change of momentum[25]. In either case, we need to know the change in velocity, $\Delta\underline{v}$, which is equal to the final velocity minus the inital velocity. In vector form, this is equivalent to **adding** the final velocity to a vector which is minus the initial velocity (and so has the same magnitude as the initial velocity but is in the opposite direction). This, with several useful angles marked, is shown in Figure 2.37.

Figure 2.37 Finding the change in velocity of the ball.

The angle γ inside this triangle would be useful for calculating the magnitude of the change in velocity. It can be written in terms of the angles given in the question by using the fact that all angles at a point (in a plane) must sum to $360°$, as in Figure 2.38.

$$\gamma = 360° - \alpha - 90° - (90° - \beta)$$
$$= 180° - \alpha + \beta$$

Figure 2.38 Finding the angle γ.

Figure 2.39 shows how this can then be used along with the cosine rule to work out the magnitude of the change in velocity.

As stated earlier, Newton's Second Law can be used to work out the magnitude of the average force on the ball in terms of the magnitude of the change in velocity

---

[25] You may not have come across this in your classes yet. If not, can you show that, for an object of constant mass, this is equivalent to the equation $\underline{F} = m\underline{a}$?

Magnitude of Δv :

$$(\Delta v)^2 = u^2 + v^2 - 2uv\cos\gamma$$

Figure 2.39 Finding the magnitude of the change in velocity.

and the time taken for the collision, as shown in Figure 2.40[26].

From Newton's 2nd Law the magnitude of the force on the ball is given by

$$F_b = \frac{\Delta p}{t} = \frac{m\Delta v}{t}$$

Figure 2.40. Using Newton's Second Law to find the magnitude of the force on the ball.

However, the question asks for the magnitude of the force on the player, not on the ball. We can use Newton's Third Law to show that this is the same as the magnitude of the force on the ball, as shown in Figure 2.41.

Using Newton's 3rd Law the magnitude of the force on the player's head is equal to the magnitude of the force on the ball

$$F_h = F_b = \frac{m\Delta v}{t}$$

Figure 2.41. Using Newton's Third Law to find the magnitude of the force on the player.

---

[26]This is often expressed as **Impulse** $= \Delta \underline{p} = \underline{F}_{avg}\Delta t$

Finally, we can substitute our earlier value for the magnitude of the change in velocity (as shown in Figure 2.39) into this equation to give our final answer in Figure 2.42.

$$\text{Substituting in } \Delta v \text{ from earlier}$$

$$F_h = \frac{m}{t}\sqrt{u^2 + v^2 - 2uv \cos \gamma}$$

$$\text{Substituting in for } \gamma$$

$$F_h = \frac{m}{t}\sqrt{u^2 + v^2 - 2uv \cos\left(180° - \alpha + \beta\right)}$$

Figure 2.42 Substituting in to solve the problem.

Extra things to think about

Could you modify this answer to describe the force required in terms of the angle by which the ball's direction changes? How would the situation differ if the ball just "glanced" off the top of the player's head? What sort of "headers" require football players to exert less force?

2.9   Moments - Balancing an Unknown Mass

Question

> A light beam sits on top of a pivot. A person of unknown weight sits a distance
> x_1 away from the pivot while a person of weight F_2 sits opposite them a dis-
> tance x_2 from the pivot. If the beam is balanced, what is the magnitude of the
> force from the pivot acting on the beam in terms of x_1, x_2 and F_2?

Main skills required

This question uses the balance of moments when a system is in equilibrium but
illustrates that this is true for any point that you choose[27], not just the pivot point.

Worked Solution

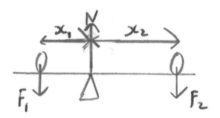

Figure 2.43. Two people balanced on a light beam, resting on a
pivot.

To draw the diagram in Figure 2.43, we can consider both people and the beam
as the system and draw the forces acting on this system[28]. As the beam is described
in the question as "light", this implies that its weight can be ignored in this question
as it is so small compared to the other forces; you'll often find the word "light" used
in this way in physics problems[29].

---

[27]This point doesn't even have to be in the body that you're looking at!

[28]The actual forces acting on the beam are the normal reaction forces from the people.
These forces will be of the same magnitude and direction as the weight of the people but are
not directly caused by the gravitational attraction of the Earth on the people - this is a very
subtle point!

[29]Although it's important not to confuse a description of an object as having little weight
with describing the massless electromagnetic waves that we use to see!

There are two unknowns on the diagram—the weight F_1 and the normal re-action force from the pivot, N. This may make you think that we need to find two equations in order to find these two unknowns, and this is a valid way to approach this question. You could take moments about the pivot point to find an equation linking the two weights and distances and then use Newton's First Law of Motion in order to link all three forces; however, as we don't actually need to find the weight F_1, we can in fact solve this problem with just one equation.

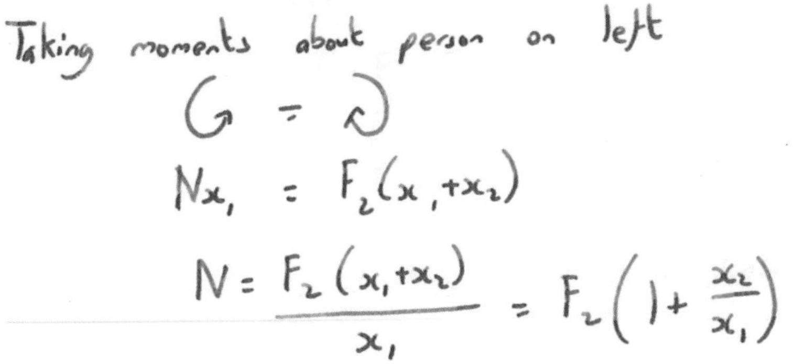

Taking moments about person on left

$$G = \curvearrowright$$

$$Nx_1 = F_2(x_1 + x_2)$$

$$N = \frac{F_2(x_1 + x_2)}{x_1} = F_2\left(1 + \frac{x_2}{x_1}\right)$$

Figure 2.44 Taking moments about an appropriate point.

Think about how we define a moment. It includes a distance, so if we can make that distance equal to zero, then we will end up with a moment of zero, and so the force causing that moment will not actually contribute to our equation at all. This means that if, instead of taking moments about the pivot point, we take moments about the person of unknown weight, we get an equation that does not involve F_1, as shown in Figure 2.44. We can do this because when an object or system[30] is in equilibrium[31] then it is the case that the total moment[32] must sum to zero when measured about any point, so we can always choose the most convenient point about which to take moments.

Extra things to think about

How would this question change if the beam was still in equilibrium but the weight of the beam had to be considered? Think about another equilibrium situation where two unknown forces act at different points and are neither parallel to or perpendicular to each other. About which point could you take moments that would mean that you didn't need to find either of those forces?

---

[30] A fancy word for a defined collection of objects.
[31] Not accelerating linearly or rotationally.
[32] Direction is important, just like for forces.

2.10   Energy - A Bike Jump

Question

> A cyclist and bicycle start stationary at the top of a slope of height h_0. They
> free-wheel down to the bottom of the slope (*i.e.*, no pedalling) and then launch
> off a ramp, reaching a height h above the ground. Assuming no friction, what
> is the speed of the bike at the bottom of the slope and at the top of the bike's
> jump? Ignore the rotation of the bicycle's wheels.

Main skills required

This question uses conservation of energy, specifically kinetic energy and gravitational potential energy measured near the Earth's surface. You also need to avoid stalling by thinking that there is information missing; the angle of the ramp is not given here, but you need to be able to deal with this and continue with the problem.

Worked Solution

In problems like this where we only need to consider the properties of an object at a few key points and don't care about the detail of the motion in between those points, we can use a method based on conservation of energy to solve the problem. In this case, this removes the need to deal with the changing direction of the velocity of the bike and rider as they travel down the slope which would involve some complicated mathematics to work out the changing acceleration and the changing normal reaction force acting from the slope[33].

As usual, the first step is to represent all of the information given in a diagram like that in Figure 2.45. This allows us to define some of the symbols that we are going to be using in our calculations[34]. Notice that at the maximum and minimum heights that are reached, the velocity of the bike and rider must have no vertical component of velocity—it is entirely horizontal.

We can then *conserve energy between two different positions*, which is the same as *conserving energy between two different times*. First, we'll conserve energy between when the bike is at the top of the slope and when it's at the bottom. As the bike is

---

[33]Think carefully about what component(s) of these forces would cause a change in speed—although they'd all contribute to a change in velocity, this could include a change in direction rather than speed.

[34]As in previous questions, I've chosen symbols so that I can easily work out what they refer to. You may want to use different symbols, but you should still think about how the letter(s) chosen help to represent a particular value.

Figure 2.45 The bike at several key points during its motion.

initially stationary, it will have had no kinetic energy, only gravitational potential energy. Conversely, at the bottom of the slope, the bike will have some kinetic energy but will have zero gravitational potential energy (as we have defined GPE to be zero at the Earth's surface—it's important to note that this is not the only possible definition, but we've chosen this one in particular because it makes our calculation easier). Figure 2.46 shows how setting the total energies at these two positions to be equal (conservation of energy) allows us to find an expression for the speed of the bike at the bottom of the slope.

$$\text{Energy at top} = \text{Energy at bottom}$$
$$\text{GPE at top} = \text{KE at bottom}$$
$$mgh_0 = \frac{1}{2}mu^2$$
$$2gh_0 = u^2$$
$$u = \sqrt{2gh_0}$$

Figure 2.46. Conserving energy between the times when the bike is at the top and bottom of the ramp.

In the same way, we can equate the total energy of the bike and rider at the top of the slope to the total energy at the highest point in the jump. Note that, as

shown in Figure 2.47, we need to think about both the kinetic and potential energy
of the bike and rider during the jump[35].

$$\text{Energy at top} = \text{Energy in mid-air}$$
$$\text{GPE at top} = \text{KE in mid-air} + \text{GPE in mid-air}$$
$$mgh_o = \tfrac{1}{2}mv^2 + mgh$$
$$gh_o - gh = \tfrac{1}{2}v^2$$
$$g(h_o - h) = \frac{v^2}{2}$$

$$\boxed{v = \sqrt{2g(h_o - h)}}$$

Figure 2.47. Conserving energy between the times when the bike is
at the top of the ramp and the highest point during the jump.

Another way of thinking about all of this is that the gravitational potential energy **lost** by the bicycle, compared to its initial position, is equal to the kinetic energy **gained** by the bicycle over the same time. Be careful with your signs if you
work through the problem this way—don't confuse the **change** in gravitational potential energy (which would be negative from the top of the slope to the bottom of
the slope) with the **loss** in gravitational potential energy (which would be positive
from the top of the slope to the bottom of the slope).

Extra things to think about

Imagine that you are now given the energy lost to friction along the downward
slope, and you are also told the energy lost to friction during the motion up the
ramp. How would this affect your calculations?

---

[35] Make sure that you are consistent with where GPE has been defined to be zero.

2.11   Circular Motion - Time Taken Round a Corner

Question

A corner on a racetrack has a radius of curvature of R and subtends an angle of θ. A racecar of mass m is driven around the track. If the maximum frictional force between the car and the track is F, what is the shortest time needed for the car to go around the corner?

Main skills required

This question requires knowledge of the acceleration, and hence centripetal force (calculated using Newton's Second Law of Motion), required for circular motion. It also requires various terms from the question to be "interpreted" to set up the situation in the first place.

Worked Solution

Several of the terms in this problem may be new to you, so the first step towards solving this problem is coming to understand what those terms mean. To define the "radius of curvature" of a curve, the curve should be treated as if it were just part of a complete circle. The radius of curvature of the curve is just the radius of this imagined complete circle. This means that tighter curves have a smaller radius of curvature. For the purposes of this question, we can treat the car going around this section of curve exactly the same as if it were going around in a circle.

The "angle subtended", shown on Figure 2.48, allows us to work out the length of this section of track. If lines were drawn from the ends of the curve to the centre of the imaginary circle (these will be of length r, the radius of curvature of the track), then the angle subtended by the circle is the angle between those two lines. We've also defined the length of the curve as l and the speed of the car travelling around it as v on this diagram, as well as showing the mass of the car. The frictional force described in the question is drawn acting towards the centre of the imaginary circle, in order to provide the centripetal force required to keep the car on the track.

As the centripetal force is provided by the frictional force F, we can equate this to the equation for the required centripetal force, as shown in Figure 2.49. This derives from the acceleration of an object moving in a circle with a constant speed, $a = \frac{v^2}{r}$, combined with Newton's Second Law of Motion $F = ma$ to give the required resultant force which will keep an object moving in a circle with a constant speed. We can then rearrange this to find the speed that the racecar must have to be moving in a circle if the frictional force is F. For the maximum speed around the corner, this frictional force is the maximum frictional force given in the question.

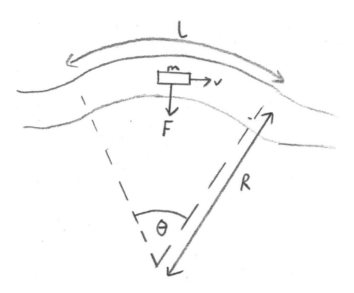

Figure 2.48. The racecar travelling around the bend, with the centripetal force required added.

In order to calculate the time taken for the car to pass through this curve, we could have instead calculated the angular speed (angle per second, ω) from the centripetal force ($F_{\text{centripetal}} = m\omega^2 r$), but we've stuck with linear speed. This means that we need to find the length of the curve. If the angle θ is given in radians, then the arc length can be calculated as shown in Figure 2.50. If the angle was given in degrees, it would first need to be converted into radians for this equation to make sense.

$$\text{Centripetal force provided by } F$$

$$F = \frac{mv^2}{R}$$

$$v = \sqrt{\frac{RF}{m}}$$

Figure 2.49. Calculating the maximum speed that the racecar can have around the bend for a maximum frictional force F.

Length of curve $L = R\theta$

Constant v so time to drive around curve:

$$t = \frac{L}{v}$$

Figure 2.50. Using an angle in radians to calculate the distance around the curve and then the time taken to travel around it.

At this point, all of the information has been extracted from the question, and our values for l and v can be substituted into the equation given in Figure 2.50 and rearranged as in Figure 2.51 to give the answer.

Substituting in expression for L & v:

$$t = \frac{R\theta}{\sqrt{\frac{RF}{m}}} = R\theta\sqrt{\frac{m}{RF}} = \theta\sqrt{\frac{Rm}{F}}$$

Figure 2.51. Combining the previous information to solve for the time taken for the racecar to travel around the bend.

Extra things to think about

From the equation obtained, we can see that it takes longer to go around wider bends that subtend the same angle, as the effect of the curve being longer outweighs the increased speed of travel possible around the corner. How would this equation change if the length of the curve was fixed rather than the angle that it subtended?

2.12   Circular Motion - With Changing Speed

Question

> A rollercoaster car of mass m, initially travelling horizontally with a speed u, enters a frictionless[a] vertical circular loop at the bottom. If the loop has a radius of r, what is the magnitude of the centripetal force acting on the rollercoaster car when it has travelled around the loop an angle of θ from the bottom?
>
> _____
> [a]Again, an oversimplification of a real situation. Once you've solved this problem, why not think about how friction would affect your answer?

Main skills required

This question requires conservation of energy (kinetic energy and gravitational potential energy near to the Earth's surface) and the acceleration needed for circular motion, as well as Newton's 2nd Law of motion and some trigonometry.

Worked Solution

In this problem we're given the initial situation (the rollercoaster car travelling horizontally) and asked to find a solution for the car for any angle θ. This means that we can draw a single situation on our diagram, but we have to make sure that we keep θ in our equations and see that our solution works for any possible value of θ around the loop (from 0 to 2π radians). One possible way to draw the diagram is shown in Figure 2.52

In Figure 2.52 we can define the height of the rollercoaster car above its initial position as h because we'll need to know this to calculate the gravitational potential energy of the car. This height was found by noticing that the radius of the loop, measured from the centre to the base of the loop, is made up of the height of the car plus one side of the triangle shown using dashed lines. It is important to check that this equation works even for angles of θ greater than $\pi/2$ radians ($90\,°$). If $\theta = \pi$ rad, then $\cos\theta = -1$, giving $h = 2r$ as required. By the symmetry of the cosine function about the angle of π radians, this equation also works for larger angles.

Although you may have come across the equation for the centripetal force required for movement in a circle at a constant speed, you may not have realised that **if you take one particular instant during circular motion, you can still use this equation**. The change in speed is due to the tangential component of the acceleration, not the radial acceleration, which is responsible for keeping the object moving in a circle.

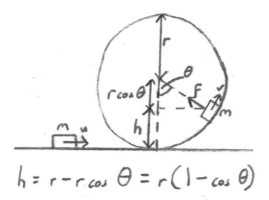

$$h = r - r\cos\theta = r(1 - \cos\theta)$$

Figure 2.52. The rollercoaster car before it enters the loop, and at an arbitrary position an angle θ around the loop from the bottom. The height at this angle is also calculated by trigonometry.

In order to use this equation, we need to know the speed of the rollercoaster car at this particular instance, which we don't yet have. This can be calculated using conservation of energy, as shown in Figure 2.53. Conservation of energy is a sensible method to use here because we don't care about the time taken to reach this height h. Also, we don't strictly need to find v, as it appears in the centripetal force equation as v^2, so we can stop once we've obtained an expression for v^2.

Conservation of energy:

$$\text{Energy at bottom} = \text{Energy at height } h$$
$$\text{KE at bottom} = (\text{KE} + \text{GPE}) \text{ at height } h$$
$$\tfrac{1}{2}mu^2 = \tfrac{1}{2}mv^2 + mgh$$
$$u^2 = v^2 + 2gh$$
$$v^2 = u^2 - 2gh$$

Figure 2.53. Conserving energy between the two positions shown in the previous diagram.

Now that we have equations for both v^2 and h, we can find the required centripetal force with a couple of substitutions, as shown in Figure 2.54.

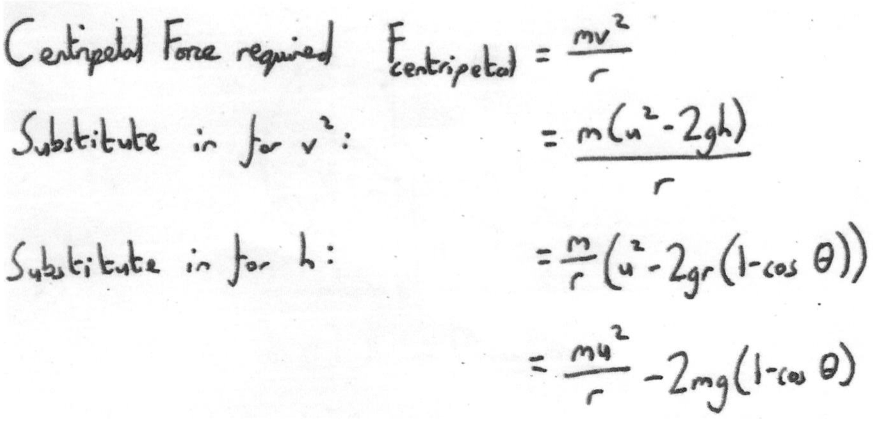

Centripetal Force required $F_{centripetal} = \frac{mv^2}{r}$

Substitute in for v^2: $= \frac{m(u^2 - 2gh)}{r}$

Substitute in for h: $= \frac{m}{r}\left(u^2 - 2gr(1-\cos\theta)\right)$

                                            $= \frac{mu^2}{r} - 2mg(1-\cos\theta)$

Figure 2.54. Substituting in to find the magnitude of the centripetal force at any position around the loop.

Extra things to think about

Depending on the position of the car on the loop, the centripetal force could be made up of a combination of the weight of the car and the normal reaction force on the track. How would you use your answer from the problem given if you were instead asked to calculate the normal reaction force acting on the car at various points on the track?

The car is changing speed. Which force acting on it causes this change in speed, and specifically which component of that force?

2.13   SHM - Velocity of a Mass on a Spring

Question

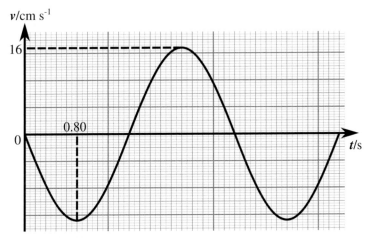

Figure 2.55 Velocity-time graph of an unknown mass on a spring.

An unknown mass is placed on a frictionless table and attached to a spring which is attached to a fixed point. The spring is stretched, and then the mass is released. Figure 2.55 shows the velocity of the mass after this time. The spring constant of the spring is $k = 85\,\mathrm{N\,m^{-1}}$.

a Find:

   i the time period of the simple harmonic motion.

   ii the maximum displacement of the mass from the equilibrium position.

   iii the mass of the block.

b From the graph, estimate the maximum acceleration of the mass and compare this to a calculated value.

Main skills required

This question requires the ability to read data from a graph and use it to find other values based on your knowledge of simple harmonic motion. It also uses conservation of energy.

Worked Solution

There are a few pieces of information on the graph and one given in the question, but these are sufficient for us to completely understand this system if we use them correctly. The first part of this question eases us into this, as it only requires one piece of information. We can read off a time directly from the graph, but this is not the time period of the motion. Instead, we have to notice that it's one quarter of the time period, which means that the time period can be calculated, as shown in Figure 2.56.

$$\text{From graph,} \quad \frac{T}{4} = 0.80\,s$$
$$\underline{T = 3.2s}$$

Figure 2.56. Calculating the time period from the values given on the graph.

As that was simply reading values from the graph provided, it was possible to answer the first part of the question without drawing a diagram of the situation. With the next part of the question, there are a few more things happening, so a diagram is needed—Figure 2.57. Notice in particular how I've been very explicit about describing the "zero" point here—if you don't give this, it can be hard to work out what is meant by displacements, *etc.* It can be convenient for this to be the point where the resultant force is zero, and the speed of the mass is at a maximum.

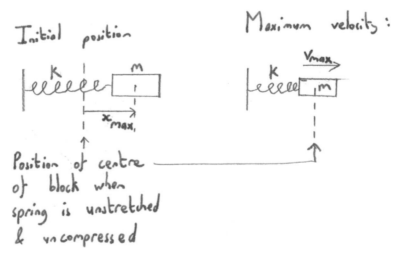

Figure 2.57. The mass and spring in its intial position, and at the point when the mass has its maximum velocity.

While in the first part of the question we explicitly cared about the time, we are now looking at a situation where we are just trying to find a displacement at one particular point and we don't need to know the time at which it occurs. This means that we can use conservation of energy, as in Figure 2.58. I noticed that the equation contained the mass, which we don't have, but as a mass on a spring is a common system, I know that this mass also appears in the equation for the angular frequency of the motion[36] and hence the period.

Using conservation of energy:

Initial elastic potential energy = Maximum kinetic energy

$$\tfrac{1}{2}k\,x_{max}^2 = \tfrac{1}{2}m\,v_{max}^2$$

$$x_{max}^2 = \frac{m\,v_{max}^2}{k}$$

But using $\omega^2 = \frac{k}{m}$

$$x_{max}^2 = \frac{v_{max}^2}{\omega^2}$$

Figure 2.58. Conserving energy between the two situations shown in Figure 2.57.

Although we don't know the angular frequency, we can calculate it from the time period I found in the first part of this question. This means that we can substitute the equation for ω in Figure 2.59 and get an equation for the maximum displacement of the mass containing only values we have obtained from the graph.

---

[36]I'm being a little naughty here and should really derive this from the system. Check that you could derive this by using the form of the equation of motion for SHM: $a = -\omega^2 x$.

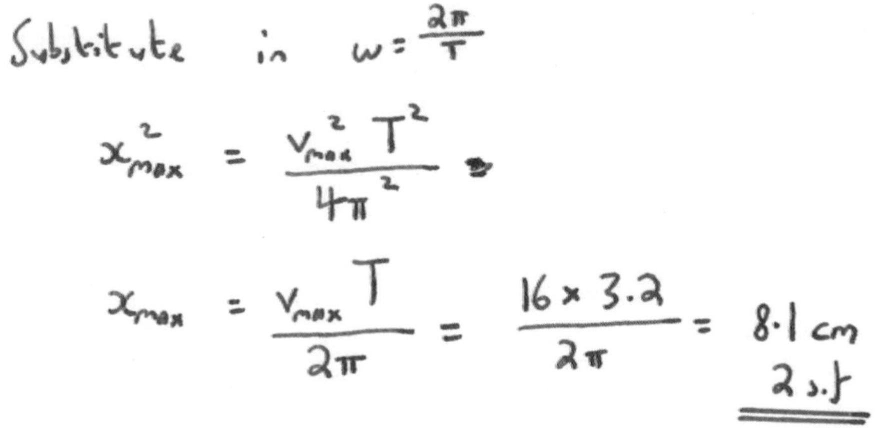

Substitute   in $\omega = \frac{2\pi}{T}$

$$x_{max}^2 = \frac{v_{max}^2 \, T^2}{4\pi^2}$$

$$x_{max} = \frac{v_{max} \, T}{2\pi} = \frac{16 \times 3.2}{2\pi} = \underline{\underline{8.1 \text{ cm}}}$$
$$\phantom{x_{max} = \frac{v_{max} \, T}{2\pi} = \frac{16 \times 3.2}{2\pi} = } 2 \text{ s.f}$$

Figure 2.59. Substituting in for the time period, using the definition of angular frequency of oscillation.

Finding the mass means taking a step back using the equation for the angular frequency that we used previously. A few substitutions, and the answer is obtained in Figure 2.60.

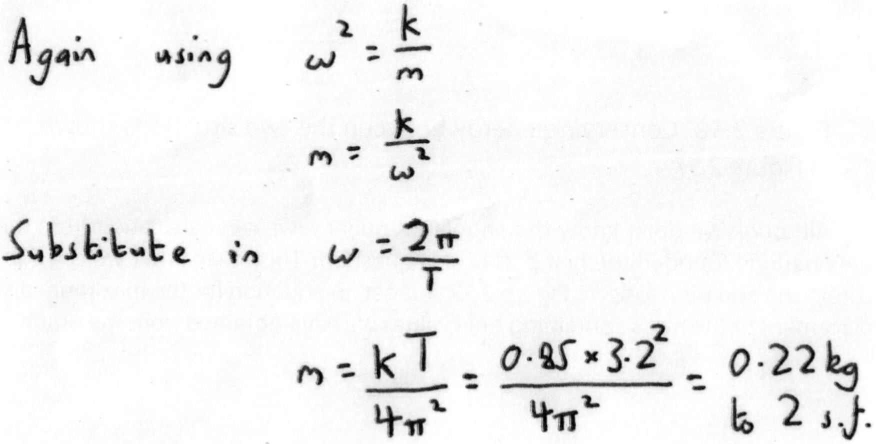

Again using $\omega^2 = \frac{k}{m}$

$$m = \frac{k}{\omega^2}$$

Substitute in $\omega = \frac{2\pi}{T}$

$$m = \frac{kT^2}{4\pi^2} = \frac{0.85 \times 3.2^2}{4\pi^2} = \underline{\underline{0.22 \text{ kg}}}$$
$$\phantom{m = \frac{kT^2}{4\pi^2} = \frac{0.85 \times 3.2^2}{4\pi^2} = } \text{to 2 s.f.}$$

Figure 2.60 Solving for the mass.

The next part of the question is there to demonstrate once again how everything has to end up being consistent. First, in Figure 2.61, we can use the definition of acceleration and a tangent drawn to the point of maximum acceleration[37] to find the acceleration.

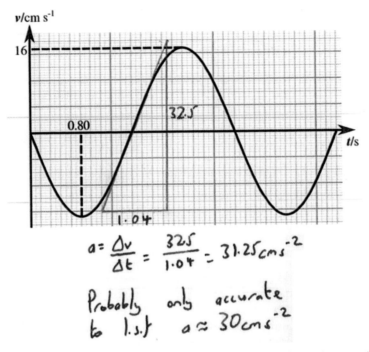

$$a = \frac{\Delta v}{\Delta t} = \frac{32.5}{1.04} = 31.25 \, cms^{-2}$$

Probably only accurate
to 1.s.f. $a \approx 30 cms^{-2}$

Figure 2.61 Estimating the maximum accleration from the graph.

In Figure 2.62 I've again skipped to a known result for SHM[38], that the maximum acceleration is given by $a_{max} = \omega v_{max}$, and substituted in values from the graph—I've deliberately not used calculated values so as to avoid any rounding errors. As this has come from values explicitly given to 2 s.f. rather than trying to read points from the graph, this answer is likely to be more precise than that calculated in Figure 2.61.

---

[37] Can you justify what the maximum acceleration is when the velocity is zero? What would the displacement be here?

[38] Again you should try to derive this, starting by finding the maximum force on the mass.

$$a_{max} = \omega v_{max}$$

$$= \frac{2\pi v_{max}}{T}$$

$$= \frac{2\pi \times 16}{3.2}$$

$$= 31.42 \, cms^{-2} = 31 \, cms^{-2}$$
$$\text{to } 2 \, s.f.$$

Figure 2.62. Calculating the maximum acceleration from the known maximum velocity and time period.

Extra things to think about

How would this situation change if the mass was instead hung vertically from the same spring? If you didn't know the spring constant of the spring but did know the mass that you attached to it, would you be able to solve this problem? Is there another experiment you'd need? Another common system exhibiting simple harmonic motion is a simple pendulum consisting of a mass on a light string. Could you answer a similar question for that system? What would be different and what would remain the same?

Circuits

3.1 Circuits - Potential Divider

Question

Two resistors of resistances R_1 and R_2 are connected in series. A potential V_{in} is applied across this combination of resistors. Find:

i the potential difference across each resistor.

ii the resistance of a single resistor which could be used to replace this set of two resistors without changing the current flowing through this circuit.

Main skills required

This is an example of a circuit problem where there are equations that are often taught to students which would allow them to solve the problem quickly; however, working through the problem using more fundamental rules is useful because these fundamental rules are more widely applicable. The solution to this problem will use Kirchhoff's Voltage Law, Kirchhoff's Current Law, and the relationship $V = IR$ for a component of constant resistance R (Ohm's Law).

Worked Solution

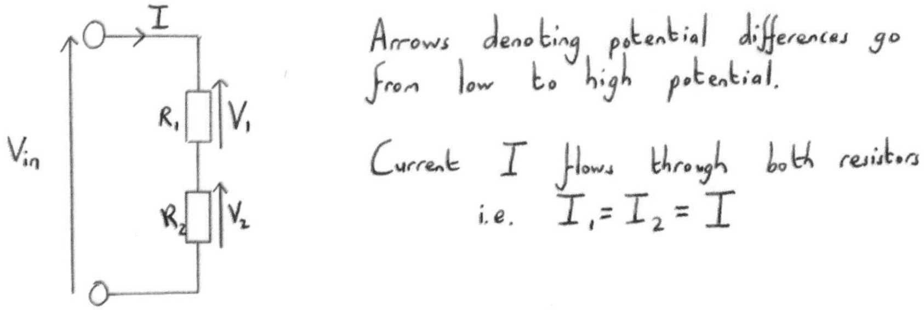

I
R1
V1
Vin
R2
V2
Arrows denoting potential differences go from low to high potential.
Current I flows through both resistors
i.e. I1 = I2 = I

Figure 3.1. Circuit diagram for the potential diagram, with useful potential differences and current labelled.

As usual, the first thing to do is to convert the information in the text into a diagram, Figure 3.1. As well as using the symbols given in the question, I've defined symbols for the current through the circuit and the potential differences across each resistor. We only need to define one current because, from Kirchhoff's Current Law, the current going into each resistor is the current going out of the resistor, so the same current flows around the entire circuit.

The reason that I've drawn directional arrows for my potential differences is because it makes it easier to apply Kirchhoff's Voltage Law—the sum of all of the potential changes around a complete loop of the circuit must be zero[1]. I've drawn the arrows in the direction of increasing potential[2], so as you move around the loop, moving in the direction of an arrow will give you a positive potential difference while moving in the opposite direction to an arrow will give you a negative potential difference. Remember that V_{in} is an emf but the resistors dissipate energy and so have a potential drop across them when you move in the direction of the current. I've then used Kirchhoff's Voltage Law in Figure 3.2.

From Kirchhoff's Voltage Law, going clockwise.

$$V_{in} - V_1 - V_2 = 0$$
$$V_{in} = V_1 + V_2 \qquad ①$$

For resistors

$$V_1 = IR_1 \qquad ②$$
$$V_2 = IR_2 \qquad ③$$

Figure 3.2 Using Kirchhoff's Voltage Law and $V = IR$.

In Figure 3.2 I've also written out the two equations linking the current and potential across each resistor - applying $V = IR$ for each resistor. I've labelled each equation by drawing a circled number to make them easier to refer to in the rest of our working. All of the physics has been done here—the rest is just algebraic manipulation, as shown in Figure 3.3.

---

[1]This is the same as saying that the emf. is equal to the potential lost in the rest of the circuit.

[2]Which is in the opposite direction to the current through a resistor.

Substituting ② & ③ into ①

$$V_{in} = IR_1 + IR_2 = I(R_1+R_2)$$

$$I = \frac{V_{in}}{(R_1+R_2)} \qquad ④$$

Substituting ④ into ② & ③

$$V_1 = \frac{V_{in}R_1}{(R_1+R_2)}$$

$$V_2 = \frac{V_{in}R_2}{(R_1+R_2)}$$

Figure 3.3 Algebraic manipulation to find our solutions.

As expected, combining and rearranging these equations gives us the potential divider equations you may have seen before. Some similar manipulation, as shown in Figure3.4 will allow us to work out the "equivalent resistor" that could be used in the same circuit to give the same current flow when a potential V_{in} is applied, giving an equation which is commonly taught as the method to combine resistors in series.

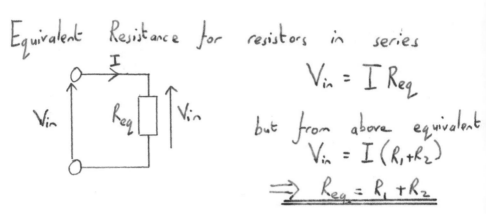

Figure 3.4. Modelling the two resistors in series as a single "equivalent" resistor.

Extra things to think about

Replace the resistor R_1 in this question with two more resistors of resistances R_3 and R_4 connected in parallel. Can you use Kirchhoff's Laws to work out the current through each resistor, without using any rules you know for "adding resistances"?

3.2   Circuits - Combining Resistors

Question

What is the current from the cell in the circuit shown in Figure 3.5?

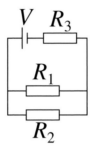

Figure 3.5 An arrangement of three resistors and a cell.

Main skills required

This is another example of a circuit problem where there are equations that are often taught to students which would allow them to solve the problem quickly, but working through the problem using more fundamental rules is useful because these fundamental rules are more widely applicable. The solution to this problem will use Kirchhoff's Voltage Law, Kirchhoff's Current Law, and the relationship $V = IR$ for a component of constant resistance R (Ohm's Law).

Worked Solution

Although we're given a circuit diagram in this question, it's still useful to redraw it so that you can add useful additional information like defining the currents through the circuit as I've done in Figure 3.6. Some people also find it easier to change what the circuit looks like so that you can more easily see that the resistor R_3 is in series with the combination of resistors R_1 and R_2, but if you do reshape the circuit, make sure that you don't change how the components are connected.

In order to work towards finding the unknown current from the cell, we're going to have to start relating other unknown quantities so that we don't need to know all of them individually[3]. Another way of thinking about this is that we're setting

---

[3] Although we can work back after we've solved the problem to find all of these.

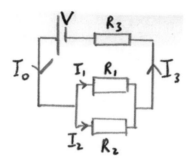

Figure 3.6. The circuit diagram redrawn and labelled with useul currents.

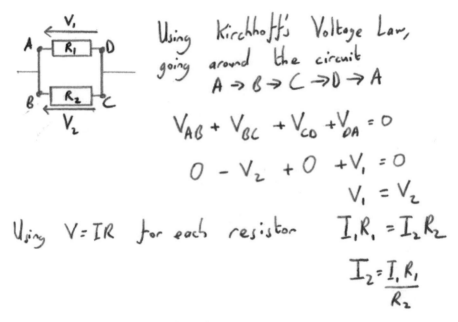

Using Kirchhoff's Voltage Law, going around the circuit

$$A \rightarrow B \rightarrow C \rightarrow D \rightarrow A$$

$$V_{AB} + V_{BC} + V_{CD} + V_{DA} = 0$$

$$0 - V_2 + 0 + V_1 = 0$$

$$V_1 = V_2$$

Using $V = IR$ for each resistor

$$I_1 R_1 = I_2 R_2$$

$$I_2 = \frac{I_1 R_1}{R_2}$$

Figure 3.7. Applying Kirchhoff's Voltage Law to the two resistors connected in parallel.

up enough simultaneous equations to find all the unknown variables[4], but we're doing it in such a way that we can focus on the one we're asked for in the question, I_0. I've started this by looking at the resistors in parallel. I've redrawn this part of the circuit in Figure 3.7 with a few points and potential differences defined. I've drawn

---

[4]The currents I_0, I_1, I_2 and I_3, as well as potentials across each resistor V_1, V_2 and V_3.

the two potential differences with arrows going from lower to higher potentials; for a resistor we know that the current flows from high to low potential, so each arrow is in the opposite direction to the current flow we defined in Figure 3.6.

The next step is to use Kirchhoff's Voltage Law, taking care to choose a direction around the circuit, in this case anti-clockwise from A to B to C to D to A. Using this direction gives us a minus sign for V_{BC} because the potential gain is in the opposite direction to the way we're looking around the circuit—the potential difference from B to C is a loss in potential. Also, as the wires are treated as perfect conductors[5], there is no change in potential along the wires, giving $V_{AB} = V_{CD} = 0$. We can then use the equation $V = IR$ to get an equation which, while still containing some unknowns (I_1 and I_2), at least also contains the known[6] quantities R_1 and R_2. We can rearrange this equation to get an equation for one of our unknowns in terms of the other: I've chosen to write I_2 in terms of I_1. This means that in other equations we'll now be able to eliminate I_2, so we have one fewer unknown to worry about.

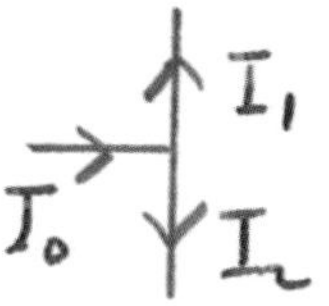

$$I_0 = I_1 + I_2$$

Substituting in for I_2

$$I_0 = I_1 + \frac{I_1 R_1}{R_2}$$

$$= I_1\left(1 + \frac{R_1}{R_2}\right)$$

Figure 3.8 Obtaining an equation for the current I_0.

A possible next step is to use Kirchhoff's Current Law to work out the relationship between the answer that we want, I_0, and the current we've just written an equation for. The junction just to the left of the two resistors in parallel is shown in Figure 3.8 with the currents going in and out of that junction labelled. By substituting in the equation we got for I_2, we can get an equation for I_0 in terms of I_1, R_1

[5] An approximation, but a good one if the wires have very little resistance compared to any of the resistors, which would be the case for most wires.

[6] Even though we don't have a value for them, they are given in the question and therefore assumed to be known.

and R_2. From looking at the current going into and out of the cell in Figure 3.6, we can also use Kirchhoff's Current Law to show that $I_3 = I_0$.

There are a few different ways that we could now continue to solve the problem[7]. In Figure 3.9 I've redrawn the entire circuit to clearly show the loop including resistors R_1 and R_3 that we're now going to use to apply Kirchhoff's Voltage Law. I've chosen this one as we've already got an equation linking the two unknowns I_0 and I_1, so another equation linking them will be enough for us to eliminate I_1 and find I_0, the target of the question.

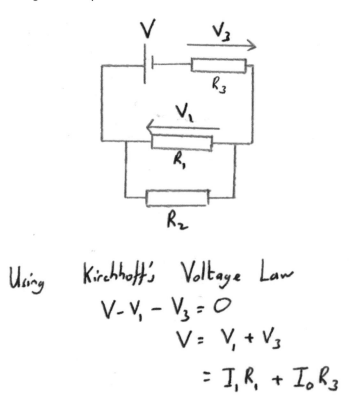

Figure 3.9 Considering a loop around the circuit including the cell.

From there, it's just manipulation of the two equations we've obtained to find the answer, as shown in Figure 3.10.

---

[7] Actually, we could have done most of these steps in whatever order we wanted. The key to these kinds of questions is properly applying Kirchhoff's laws and using $V = IR$ for resistors.

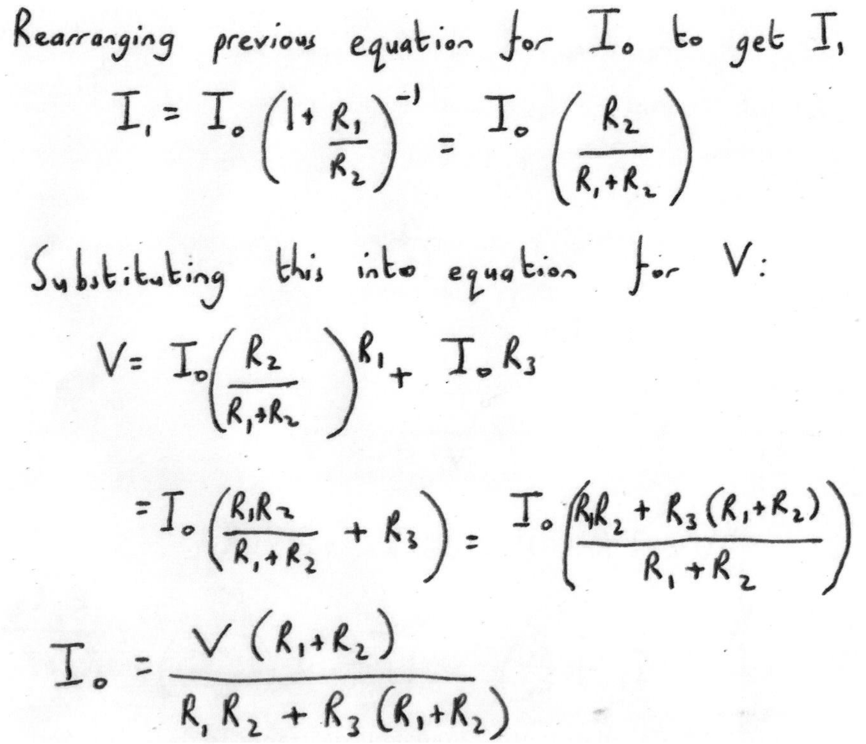

Rearranging previous equation for I_o to get I_1

$$I_1 = I_o \left(1 + \frac{R_1}{R_2}\right)^{-1} = I_o \left(\frac{R_2}{R_1 + R_2}\right)$$

Substituting this into equation for V:

$$V = I_o \left(\frac{R_2}{R_1 + R_2}\right) R_1 + I_o R_3$$

$$= I_o \left(\frac{R_1 R_2}{R_1 + R_2} + R_3\right) = I_o \left(\frac{R_1 R_2 + R_3(R_1 + R_2)}{R_1 + R_2}\right)$$

$$I_o = \frac{V(R_1 + R_2)}{R_1 R_2 + R_3(R_1 + R_2)}$$

Figure 3.10 Substituting in to solve the problem.

You might recognise the form of the "equivalent resistance" in the denominator of the equation we've obtained for the current from the cell. This highlights another way of solving this problem, in which the resistors R_1 and R_2 are first combined into an equivalent resistor R_{12}. This combination is shown in Figure 3.11.

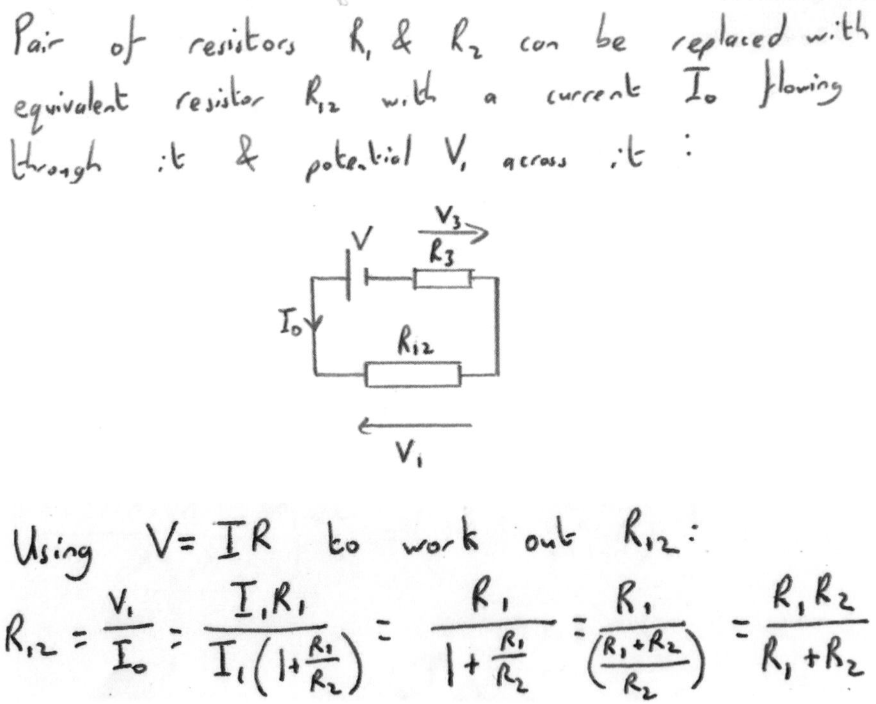

Pair of resistors R_1 & R_2 can be replaced with equivalent resistor R_{12} with a current I_o flowing through it & potential V_1 across it :

Using $V = IR$ to work out R_{12}:

$$R_{12} = \frac{V_1}{I_o} = \frac{I_1 R_1}{I_1\left(1+\frac{R_1}{R_2}\right)} = \frac{R_1}{1+\frac{R_1}{R_2}} = \frac{R_1}{\left(\frac{R_1+R_2}{R_2}\right)} = \frac{R_1 R_2}{R_1 + R_2}$$

Figure 3.11. Calculating the equivalent resistance to the pair of resistors R_1 and R_2.

This "equivlent resistor" can then be used along with Kirchhoff's Voltage Law to solve the problem, as shown in Figure 3.12.

This solution can also be written as the potential provided by the cell divided by an equivalent resistance R_{eq} for the combination of the three resistors, obtained by first combining R_1 and R_2 in parallel to get an equivalent resistor R_{12}, then combining this in series with the resistor R_3—which would have been another method of solving this problem. You may be more familiar with this method, but it's good to be able to solve problems using Kirchhoff's Laws because they give a more fundamental understanding of what is going on in a circuit and because you don't need to try to work out which resistors, or sets of resistors, are connected in parallel or in series.

Using Kirchhoff's Voltage Law

$$V - V_1 - V_3 = 0$$
$$V = V_1 + V_3$$
$$= I_o R_{12} + I_o R_3$$
$$V = I_o (R_{12} + R_3)$$

$$I_o = \frac{V}{R_{12} + R_3}$$

$$\Rightarrow I_o = \frac{V}{\left(\frac{R_1 R_2}{R_1 + R_2}\right) + R_3} = \frac{V}{R_{eq}}$$

$$\Rightarrow R_{eq} = \frac{R_1 R_2}{R_1 + R_2} + R_3 = \frac{R_1 R_2 + R_3 (R_1 + R_2)}{R_1 + R_2}$$

Figure 3.12 Solving the problem using the equivalent resistor.

Extra things to think about

Can you write down the current flowing through each resistor in terms of only the potential from the cell and the resistances of all the resistors? How would you solve this problem if another resistor was added in one of the two "parallel arms" of the circuit? What if another resistor was added in parallel with the cell?

3.3   Circuits - Maximising Power Transfer

Question

A circuit consists of a battery which provides an emf of ε with an internal resistance r connected to a load with a variable resistance.

  i For what value of load resistance will the largest power be transferred to the load, and what is this power?

  ii What power is transferred if the load has a very small or a very high resistance?

Main skills required

As well as Kirchhoff's Laws, this question uses differentiation to find a stationary point. It also demonstrates how to use an equation to find the behaviour of a system as it tends towards high and low values.

Worked Solution

Figure 3.13 shows the circuit diagram we can draw to represent this situation. We can explicity separate the emf provided by the battery and its internal resistance, which is modelled as a resistor of resistance r in series with the battery. We can also define the current through the circuit to be I, as we know that this must be the same all the way around the series circuit due to Kirchhoff's Current Law.

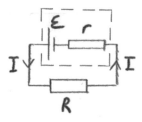

Figure 3.13. Circuit diagram with the battery represented as a separate cell and resistor.

In order to find the power dissipated in the load, which I have defined to have a resistance R, we need to know either the current through it or the potential dropped across it[8]. The easiest thing to work out is the current, as this is the same through the entire series circuit, as stated earlier. Using Kirchhoff's Voltage Law and $V = IR$ for each of the resistances in the circuit gives the current shown in Figure 3.14.

$$\text{Using Kirchhoff's Voltage Law}$$
$$\varepsilon - IR - Ir = 0$$
$$\varepsilon = I(R+r)$$
$$I = \frac{\varepsilon}{(R+r)}$$

Figure 3.14 Using Kirchhoff's Voltage Law to find the current.

This can then be substituted into an equation for the power dissipated in a resistor[9], $P = I^2R$, to give Figure 3.15.

$$\text{Power dissipated in resistor } R$$
$$P_R = I^2 R = \frac{\varepsilon^2 R}{(R+r)^2}$$

Figure 3.15 Power dissipated in resistor.

We're asked to find the maximum power transferred to this resistor, so we need to maximise this P_R. If you think about a graph of power against the resistance R of this load, at the maximum power the graph will have a slope of zero. This is equivalent to the differential of P_R with respect to R being zero, as stated in Figure 3.16. Differentiating our equation for P_R and setting this differential equal to zero gives us an equation to solve, shown in Figure 3.16. Simplifying this gives our answer, which is a result known as load matching—the greatest power will be transferred to the load when its resistance matches the internal resistance of the power supply.

---

[8]As these are linked by $V = IR$ and the power is given by $P = IV$, we only need one of V and I to calculate P.

[9]Make sure that you can derive this equation from the general equation for power in a circuit, $P = IV$.

At this point, half of the power supplied by the battery is wasted thermally inside the battery—it can only be 50% efficient.

$$\text{Maximum when } \frac{dP_R}{dR} = 0$$

$$\frac{dP_R}{dR} = \frac{\varepsilon^2}{(r+R)^2} - \frac{2\varepsilon^2 R}{(r+R)^3} = 0$$

$$1 - \frac{2R}{r+R} = 0$$

$$r + R = 2R$$

$$\underline{\underline{r = R}}$$

Figure 3.16. Differentiating the power dissipated by the resistance R to find the maximum power dissipated.

The second part of the question asks what happens at very small or high resistances. Here we're going to have to make some approximations based on the relative sizes of similar terms. In this case the equation for power given in Figure 3.15 has a term $R + r$, where r is fixed and we are looking at the behaviour as R changes. If R is very small, then this term is going to be close to just r, as shown in Figure 3.17. Substituting this in to replace the $R + r$ term in the power equation gives an equation which is linear in R, so for small enough R the graph looks approximately linear, going to zero as the load tends to zero resistance. This makes sense because, for loads with small resistances, the current will be limited by the internal resistance of the battery and so will be approximately constant, leaving R as the only changing term in $P = I^2 R$.

We can do the same for very large R[10], as in Figure 3.18. In this case r will be much smaller than R, so the term $R + r$ can be approximated as R. Substituting this

---

[10]Which I've written as tending to infinity in the diagram.

$$\text{As} \quad R \to 0 \ , \quad (R+r) \to r$$

$$P_R \to \frac{\varepsilon^2 R}{r^2} \to 0$$

Figure 3.17. Finding the power dissipated by the resistor if it has a very small resistance.

in and simplifying shows that for large R the power dissipated in the load is inversely proportional to the resistance of the load, tending towards zero as this resistance increases. This again makes sense, as the internal resistance now plays effectively no role, and nearly all of the emf is dropped across the load, and so the situation is best explained with the power equation $P = V^2/R$, which shows the inversely proportional relationship between the power and the resistance of the load for a fixed emf.

$$\text{As} \quad R \to \infty \quad (R+r) \to R$$

$$P_R \to \frac{\varepsilon^2 R}{R^2} = \frac{\varepsilon^2}{R} \to 0$$

Figure 3.18. Finding the power dissipated by the resistor if it has a very large resistance.

Thinking about how an equation behaves as a variable becomes very small or very large can be a good check once you've obtained an answer to a physics problem. Does the equation behave in the way that you expect? If not, you may have made an algebraic error—or the system may just do something unexpected! At this point you may want to check your working, just to be sure.

Extra things to think about

What would happen if you tried to find the maximum power transferred to the load by changing the emf provided? What point would you obtain instead? What other calculus could you do to show that this was not a maximum?

3.4   Capacitors - Current and Charge

Question

A circuit consists of a variable DC supply, a capacitor of capacitance $C = 860\,\mu$F, and a resistor of resistance $R = 1.3$ kΩ connected in series. The capacitor is initially uncharged with the power supply switched off.
The DC power supply is switched on and continually adjusted to maintain a constant current in the circuit. After a time $t = 3.5$ s, the power supply is supplying a voltage of $V_1 = 12$ V. What is the value of the constant current in the circuit?

Main skills required

This question uses Kirchhoff's Laws, the definition of capacitance, and the link between current and charge.

Worked Solution

For this question, I've decided to show the situation after a time t rather than at zero time so that we can include the charge on the capacitor as well as the potential given in the question. These are shown in Figure 3.19.

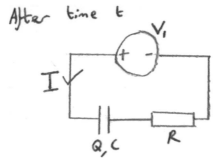

Figure 3.19. Circuit diagram for a time t after the power supply has been turned on.

Even though this question involves a capacitor, the circuit must still obey Kirchhoff's Laws. This means that the current between the components is the same everywhere, as it is in a series circuit. It also means that we can write down Kirchhoff's Voltage Law, as I have done in Figure 3.20. I've not explicitly defined V_C and V_R, but these should be clear from the context and consistency in use of symbols.

$$\text{Using Kirchhoff's Voltage Law}$$
$$V_1 - V_c - V_R = 0$$

Figure 3.20. Using Kirchhoff's Voltage Law, going anti-clockwise around the circuit in Figure 3.19.

Instead of being related to the current, the potential across a capacitor is related to the charge on it, as described in Figure 3.21. Much like $V = IR$, this can be used in Kirchhoff's Voltage Law[11].

$$\text{For a capacitor } Q = CV_c \Rightarrow V_c = \frac{Q}{C}$$
$$\text{Using this and } V = IR \text{ for resistor}$$

$$V_1 = \frac{Q}{C} + IR$$

Figure 3.21. Substituting in equations for the potential across a capacitor and a resistor.

In order to solve for the current, we need to find an equation for the unknown charge in terms of the current. Figure 3.22 shows that this can be done using the definition of current. It's important to get the sign correct here—we should make sure that the current around the circuit is defined in a way that will cause an increase in Q on the capacitor, otherwise a sign error would creep into the calculation. This then allows us to re-arrange our equation to get the current, as required.

---

[11] It's worth checking to convince yourself that the directions of the potential differences are the right way around to give the signs that I've used here. If, for example, we just had a circuit containing a charged capacitor discharging through a resistor, the potential differences across the capacitor and resistor would be in opposite directions around the circuit

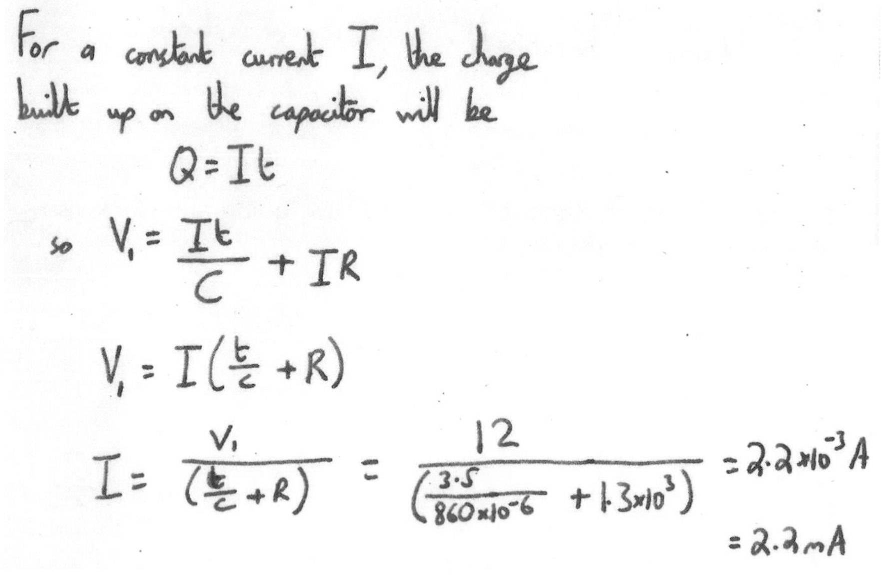

For a constant current I, the charge built up on the capacitor will be

$$Q = It$$

so $$V_1 = \frac{It}{C} + IR$$

$$V_1 = I\left(\frac{t}{C} + R\right)$$

$$I = \frac{V_1}{\left(\frac{t}{C} + R\right)} = \frac{12}{\left(\frac{3 \cdot 5}{860 \times 10^{-6}} + 1.3 \times 10^{3}\right)} = 2.2 \times 10^{-3} A$$

$$= 2.2 \, mA$$

Figure 3.22. Using the definition of a constant current to solve the problem.

Extra things to think about

How could you deal with a question involving an AC power supply so that the direction of the current changed? What would a negative charge on the capacitor represent, given that a capacitor will always have a positive charge on one side and a negative charge on the other?

Fields

Question

An electron of mass m, travelling with an initial speed u, enters a region containing a uniform electric field of constant magnitude E. The electric field points in the opposite direction to the electron's motion. After the electron has moved a distance x within the field, it will have a speed v. Find an expression for the square of this speed in terms of u, x, e, E and m.

Main skills required

This question shows how for constant acceleration it's possible to link the "before" and "after" situations either through looking at the acceleration over a distance or by a method looking at the changes in energy involved. The former method involves Newton's 2nd Law and the SUVAT equations of motion and would also be able to give information about the time taken for the motion. The latter method involves the concepts of work and kinetic energy.

Worked Solution

All of the given information should first be placed into a diagram. In this case all of the motion and forces are acting along the same line, so this is a 1-dimensional problem. The next step is to define the positive direction, which I have chosen to be to the right, in the same direction as the electron's motion. This definition will apply for all the vectors in this problem, so in Figure 4.1 the velocities of the electron $\underline{u}$ and $\underline{v}$ will both be positive, but the electric field $\underline{E}$ acts in the negative direction, and so in our calculations it will be represented as $-E$.

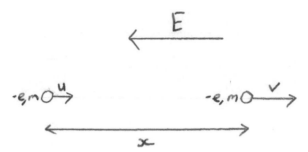

Figure 4.1. Electron before it reaches the electric field, and when it has travelled a distance x into it.

For both of the two methods we're going to look at, the force acting on the electron is needed, which we can work out using $\underline{F} = q\underline{E}$, as shown in Figure 4.2. As the charge on the electron is negative and I've drawn the electric field acting in the negative direction (acting in the opposite direction to the electron's motion as stated in the question), these both have a negative sign when substituted into the equation $\underline{F} = q\underline{E}$.

Taking to the right as positive,
force on electron

$$\underline{F} = q\underline{E} \quad \text{where } \underline{F} \text{ \& } \underline{E} \text{ are vectors}$$

$$F = (-e)(-E) = eE \text{ to the right}$$

Figure 4.2 Finding the force on the electron in Figure 4.1

For the first method, this force can then be substituted into Newton's Second Law. This is shown in Figure 4.3, with Newton's Second Law in the common form for an object with constant mass.

$$\text{Using Newton's } 2^{nd} \text{ Law}$$

$$F_{res} = ma$$

$$eE = ma$$

$$a = \frac{eE}{m}$$

Figure 4.3. Using Newton's 2nd Law to find the acceleration of the electron.

As the acceleration is constant, we can use the SUVAT equations of motion. These allow the final two variables to be calculated from a set of three. We know the initial velocity and the displacement from the question, and have now calculated the acceleration. As we want to find the square of the final velocity, v^2, we need to use the SUVAT equation that includes these four terms.

Acceleration is constant within electric field.
Using suvat equations of motion

$$s = x$$
$$u = u$$
$$v = v = ?$$
$$a = eE/m$$
$$t = ?$$

Use equation linking s, u, v and a

$$v^2 - u^2 = 2as$$

$$v^2 = u^2 + 2as$$

$$= u^2 + 2\left(\frac{eE}{m}\right)x$$

$$v^2 = u^2 + \frac{2eEx}{m}$$

Figure 4.4 Solving the problem using one of the SUVAT equations.

Alternatively, the square of the final velocity, v^2, can be found by conserving energy. At the beginning the electron has a certain amount of kinetic energy. By conservation of energy, the final kinetic energy of the electron must equal its initial kinetic energy plus any energy which is added to the system. This "added energy" is the work done by the force acting on the electron due to the electric field. It's important to make sure when calculating work done that the distance used is the component of the displacement in the direction of the force, but in this question the force and motion are both in the same direction, so this becomes the force times the distance moved.

$$\text{Alternatively, work done by force on electron}$$
$$= \text{gain in Kinetic energy}$$
$$\text{Force and displacement of electron are}$$
$$\text{in the same direction so } W = Fx$$
$$W = \tfrac{1}{2}mv^2 - \tfrac{1}{2}mu^2$$
$$Fx = \tfrac{1}{2}m(v^2 - u^2)$$
$$\frac{2Fx}{m} = v^2 - u^2$$
$$v^2 = u^2 + \frac{2Fx}{m}$$
$$v^2 = u^2 + \frac{2eEx}{m}$$

Figure 4.5. Solving the question using the work done by the electric field on the electron.

Although in this case it's just as easy to find the final situation using a method involving finding the constant acceleration, there are often situations where this is more complicated and the energy method is easier (often due to a varying force and a convenient equation to calculate a change in energy rather that using the work done). Try using both methods as you work through problems to help you get a feel for which method seems easiest for each problem.

Extra things to think about

How would this situation be different with a proton instead of an electron?

4.2   Conservation of Energy & Momentum - Two Charges

Question

Two particles, each of mass m, are separated by a distance r. One of the particles has a charge of $+q$ while the other has a charge of $+2q$. They are initially stationary and are then released. What is the speed squared of each particle when they reach a separation of $2r$?

Main skills required

This requires the use of both conservation of momentum and conservation of energy. The energy calculation requires the use of electrical potential energy.

Worked Solution

As always, we start by putting the information that we need in a diagram. Here we have two different situations at different times, so I'll represent them each with a separate diagram. The first, Figure 4.6, shows the stationary particles before they are released.

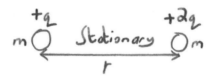

Figure 4.6 Two charged particles at rest.

The second diagram, Figure 4.7, shows the same particles after their separation has doubled. I've drawn their velocities in these directions because we know that they will have repelled each other from the starting position, as they have charges with the same sign.

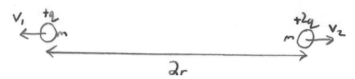

Figure 4.7. The same two charged particles as in Figure 4.6, having repelled each other.

As we're dealing with these two particles in isolation and there are no external forces, we can conserve both energy and momentum. Either could be done first, but it's sensible for us to start with momentum, as we can see from Figure 4.6 that the initial momentum was zero, and so the final momentum must also be zero[1]. Momentum is a vector quantity, so the two particles must have opposite momenta[2], and as they have the same mass, this means that their velocities must be opposite[3]. Figure 4.8 shows this more mathematically—the particles both have the same speed, and I have defined them as moving in opposite directions by drawing the arrows in opposite directions in Figure 4.7.

Conservation of momentum (Taking to the right as positive):

$$\text{Momentum after} = \text{Momentum Before}$$

$$mv_2 - mv_1 = 0$$

$$mv_2 = mv_1$$

$$v_2 = v_1$$

$\Rightarrow$ From now on use v to represent this speed

Figure 4.8. Using conservation of momentum to find the relationship between the speeds of the particles.

---

[1]Although it's not used in solving this problem, it can be worth noticing that although the two charges are different, by Newton's Third Law the magnitude of the electrostatic force acting on each charge is the same.

[2]This is the plural of momentum.

[3]Sometimes people describe this as "equal and opposite", which is short for "equal magnitude and opposite direction", but some people dislike this as without giving the full description the words "opposite" and "equal" contradict each other.

As the speeds of the two particles are the same, I've dropped the subscripts from the speeds, and I'm now using v to represent the speed of each particle[4]. We can then use this in the kinetic energy terms for conservation of energy in Figure 4.9. Be careful with the signs here—although the gravitational potential energy for the two point particles is always negative, electrical potential energy[5] can be positive or negative. Remember that you can check your final answer to see if it makes sense—if we got the signs wrong here, we would end up with an equation in which v^2 was negative, which clearly can't be the case.

$$\text{Conservation of energy:}$$

$$\text{EPE before} + \text{KE before} = \text{EPE after} + \text{KE after}$$

$$+ \frac{q(2q)}{4\pi\varepsilon_0 r} + 0 = + \frac{q(2q)}{4\pi\varepsilon_0 (2r)} + \frac{mv^2}{2} + \frac{mv^2}{2}$$

$$\frac{2q^2}{4\pi\varepsilon_0}\left(\frac{1}{r} - \frac{1}{4r}\right) = mv^2$$

$$v^2 = \frac{q^2}{2m\pi\varepsilon_0}\left(\frac{3}{4r}\right) = \frac{3q^2}{8m\pi\varepsilon_0 r}$$

Figure 4.9 Using conservation of energy to solve the problem.

Extra things to think about

How would this change if the particles had non-zero initial velocities? What if they started with different speeds to each other or had different masses? If one particle was negatively charged but they were initially moving directly away from each other, could you find the maximum distance apart that they could reach?

---

[4]We can get away with this because for kinetic energy we only need the speed, not the velocity.

[5]A common misconception is that one of the particles "has" potential energy, when really it's the potential energy of the entire system. When we think about the gravitational potential energy of an object on Earth, what we're really thinking about is the gravitational potential energy of the system containing that object and the Earth!

4.3   Circular Motion in a Magnetic Field

Question

> A particle of mass m and charge q undergoes circular motion in a magnetic
> field of magnitude B. What is the time period, T, of this motion?

Main skills required

This question asks you to construct a situation from more limited information, re-
quiring a stronger knowledge of circular motion and the force on a moving charged
particle in a magnetic field (Lorentz Force).

Worked Solution

Although there are a few quantities given in this question, and we can assume that
these will be sufficient to allow us to write an answer, there are also a few that we'll
need to add to our diagram and calculations for the intermediate steps in our calculation. We'll also need to consider the direction of the magnetic field as that's not
given in the question. To work out what we might need, we can draw a diagram like
Figure 4.10.

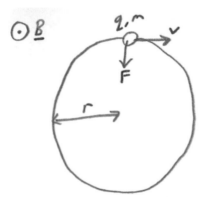

Figure 4.10. Circular motion of the charged particle, with the magnetic field acting out of the page.

You should notice that I've drawn a few things on here which are not given in
the question. As the motion is explicitly stated to be circular, I've drawn a circular
path for the particle, and so it makes sense to define the radius of this circle as r. For

circular motion we need a centripetal force, which I've labelled as F. We aren't explicilty told that the particle has a constant speed; however, **the only force that could provide the centripetal force in this situation is the magnetic force on a moving charge** which acts perpendicularly to the motion of the charge. This means that it does no work on the charge and so the kinetic energy of the charge remains constant, and so does its speed. I've therefore labelled a constant speed v on the diagram. I've also labelled the magnetic field B as acting in a **direction perpendicular to this circle** to ensure that no component of this force acts out of the plane of the circular motion (the plane of the paper in Figure 4.10), as this would mean that the motion wouldn't be circular[6].

Centripetal Acceleration $\quad a = \dfrac{v^2}{r}$

From Newton's 2nd Law, resultant force $\quad F = ma = \dfrac{mv^2}{r} \quad$ ①

Figure 4.11 Calculating the magnitude of the force needed.

As we stated previously, we need a resultant centripetal force for circular motion. The magnitude of force that is required can be calculated by combining the centripetal acceleration for circular motion at constant speed and Newton's 2nd Law of Motion, as shown in Figure 4.11.

Force provided by B field $\quad F = qvB \quad$ ②

Figure 4.12 The magnitude of the force due to the magnetic field.

We next need to consider what physical force provides the centripetal force. The magnetic force on a moving charge is always perpendicular to its motion and so will always point towards the centre of the circle drawn in Figure 4.10. This means that this force is able to provide the required centripetal force. The magnitude of this force is given in Figure 4.12—no angle appears in this equation because, as discussed earlier, the magnetic field is always perpendicular to the velocity of the particle.

---

[6]The direction here assumes that the charge on the particle is positive, and the particle moves clockwise around the circle as shown. If the charge on the particle is negative, either the direction of the magnetic field or the direction of motion of the particle (but not both) must be flipped to the opposite direction.

Equating equations ① & ② for F:

$$qvB = \frac{mv^2}{r}$$

$$v = \frac{qBr}{m}$$

Figure 4.13. Combining equations to find the speed of the particle in terms of the other quantities given, and an unknown radius r.

Equating the magnitude of this force with the required centripetal force allows us to calculate the speed of the particle which then depends on the radius of the circle as shown in Figure 4.13.

For complete circle travels a distance $2\pi r$ in a time T

$$\Rightarrow \quad v = \frac{2\pi r}{T} = \frac{qBr}{m} \quad \text{from before}$$

Rearranging for T:

$$2\pi m = qBT$$

$$T = \frac{2\pi m}{qB}$$

Figure 4.14. Combining the definition of speed with the equation for speed found in Figure 4.13.

This can then be combined with the definition of the speed and the distance around the circle in order to find the time period required by the question, as shown in Figure 4.14. You should note that the unknown speed and radius don't feature in this final answer—knowing the mass and charge of the particle as well as the magnitude of the magnetic field is enough to find this time period. This means that if we switched on a magnetic field in a situation where lots of particles were moving perpendicularly to it with different speeds, they would all move in circles with different radii but all with the same time period; faster particles would move in larger circles.

Extra things to think about

How would this situation change if the particle had a component of its velocity acting parallel to the direction of the magnetic field? What shape would describe the movement of the particles now? Which component of the velocity would you need to use in the calculations for the magnitude of the magnetic force or for the centripetal acceleration?

4.4   Gravitational Fields - Escape from Orbit

Question

A satellite of mass m is in a circular orbit around a planet of much greater mass M. If this satellite has a speed of u, how much energy must it gain in order to reach escape velocity?

Main skills required

This question requires knowledge of the centripetal force required for a circular orbit (either directly or through centripetal acceleration and Newton's Second Law), as well as an understanding of gravitational potential energy and how it relates to "escape velocity".

Worked Solution

When drawing the diagram in Figure 4.15, we have to make sure to add in a centripetal force and a radius for the orbit, as this question explicitly stated that it was going to talk about circular motion. As usual, we must make sure to use sensible letters to make it obvious what each of these are.

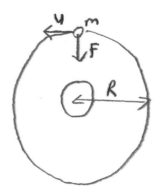

Figure 4.15 Satellite in orbit, with centripetal force shown.

The only possible force that could be providing the centripetal force is the gravitational attraction between the satellite and the planet. As the planet has a much greater mass, we can model it as stationary and take its centre to be the centre of the orbit. Equating the required centripetal force[7] with the gravitational force provided gives an equation linking the speed of the satellite and its radius, shown in Figure 4.16, which we can leave as an equation for u—we could just as easily write an equation for R.

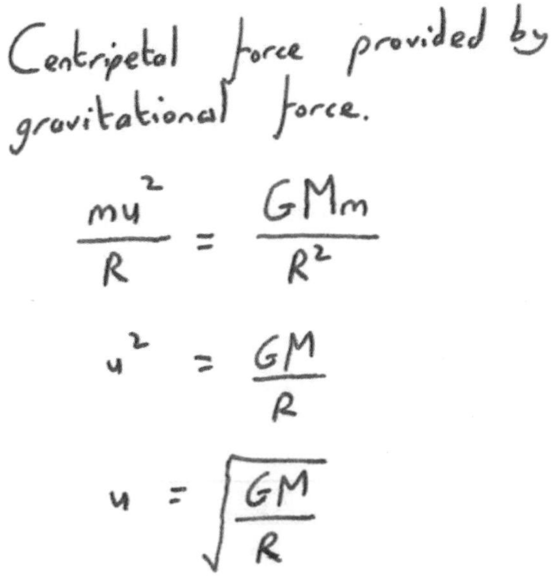

$$\text{Centripetal force provided by gravitational force.}$$

$$\frac{mu^2}{R} = \frac{GMm}{R^2}$$

$$u^2 = \frac{GM}{R}$$

$$u = \sqrt{\frac{GM}{R}}$$

Figure 4.16 Calculating the speed of the orbiting satellite.

The next step is to find the escape velocity needed, which in Figure 4.17 we get from the definition of escape velocity, that at escape velocity the sum of the kinetic and gravitational potential energies is zero. Note that for this definition to work the gravitational potential energy must be given in such a way so that it is equal to zero when the satellite is at infinity – an infinite distance away from the other mass. Although I'd normally rearrange the equation I obtained in Figure 4.17 to have v as the subject, I've noticed that as I'm going to go on to thinking in terms of energies in a further step I don't need to do this – I can keep the equation with the kinetic energy required as the subject instead.

---

[7]Make sure you can get this from the equation for centripetal acceleration, and be careful with the mass that you use.

At escape velocity v, total energy is zero.

$$KE + GPE = 0$$

$$\frac{1}{2}mv^2 - \frac{GMm}{R} = 0$$

$$\frac{1}{2}mv^2 = \frac{GMm}{R}$$

Figure 4.17. Finding an equation involving the esacpe velocity – actually finding the kinetic energy required for the object to have escape velocity.

Whenever you are talking about changes, it's important to make sure that you know which way you are going. To make sure we were doing this, in Figure 4.18 I explicitly wrote out that the additional kinetic energy needed was equal to the "final" KE (the KE at escape velocity) minus the "initial" or "current" KE. We can then substitute in for the speeds, but our answer still includes a quantity that we don't know – the radius of the orbit.

Additional KE required =

KE at escape velocity − current KE

$$\Delta KE = \frac{1}{2}mv^2 - \frac{1}{2}mu^2$$

Substituting in for $\frac{1}{2}mv^2$ and u^2 from earlier.

$$\Delta KE = \frac{GMm}{R} - \frac{m}{2}\frac{GM}{R}$$

$$= \frac{GMm}{2R}$$

Figure 4.18. Calculating the change in kinetic energy needed from the circular orbit.

In order to get rid of this unknown quantity (and a few others at the same time), we can use the equation we found in Figure 4.16. Substituting this in gives us the answer shown in Figure 4.19 that the additional kinetic energy that the satellite would need to reach escape velocity is the same amount as it already has—it would then have a total of double its current kinetic energy.

$$\text{From earlier, } \frac{GM}{R} = u^2 \quad \text{so}$$

$$\Delta KE = \frac{1}{2} m u^2$$

Figure 4.19 Substituting in for M and R to solve the problem.

Notice that even though the mass of the planet M was given as a variable, it doesn't enter into the final solution, as it is already taken into account by the combination of speed and radius needed for a circular orbit. It's unusual for questions to give you information that you don't need, but not unheard of, so if you end up in this situation, make sure to check your working, but don't assume that you must have missed something in the question.

Extra things to think about

How would you deal with a similar situation involving the orbit of two bodies of similar mass? Could you work through a similar problem involving two opposite charges?

Waves

5.1   Optics - Refraction in a Well

Question

A small torch is accidentally dropped into some water at the bottom of a well. One ray of light reaches the surface of the water a distance x from the side of the well with the horizontal component of its velocity in the direction directly towards this wall. If the angle of incidence of the beam at the surface of the water is θ, how high up the wall does the light hit the wall?
Represent the refractive index of water as n_1.

Main skills required

This question requires the use of a diagram in order to understand the situation, trigonometry to link distances on that diagram, and Snell's Law for refraction. It also requires the use of trigonometric identities to write the solution in a different form.

Worked Solution

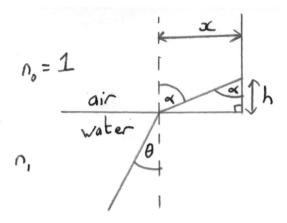

Figure 5.1. The light beam reaching the surface of the water and refracting towards the wall.

At first glance, the way this question is written may seem like it's going to be complicated to work out what it actually means; in reality the awkward phrasing is an attempt to clarify the situation for you so that you don't need to consider a fully three-dimensional situation. As the light has a horizontal component of its velocity in the direction of the wall, this means that we only need to consider that one horizontal direction – we know that if we ignore the vertical motion the light is going in the direction of the wall and not at some angle to it. This means that we've got a 2D problem, and the quantities given can be added to the diagram shown in Figure 5.1. I've also drawn in the angle α because this question must be about refraction as I've got light coming up to a boundary, and so the angle of refraction will probably be important.

Using Snell's Law

$$n_1 \sin \theta = n_0 \sin \alpha$$

$$\sin \alpha = \frac{n_1 \sin \theta}{n_0} \qquad \text{①}$$

Figure 5.2 Using Snell's Law to link the angles.

Speaking of refraction, Snell's Law allows us to link the sines of the two angles, as shown in Figure 5.2. Notice that I've labelled the resultant equation so that we can use it later. The angle α can also be used to link the distance I'm trying to find, defined as h on my diagram, to the distance given in the question—see Figure 5.3.

Using Trigonometry

$$\tan \alpha = \frac{x}{h}$$

$$\frac{\sin \alpha}{\cos \alpha} = \frac{x}{h} \qquad \text{②}$$

Figure 5.3 Linking x, h and the angle α using trigonometry.

Although we've got an equation for sin α, the trigonometry also requires cos α[1]. While we could use inverse sine[2] to find the angle and then take the cosine of that, a neater way of doing this is to write the cosine term in terms of the sine term. This means using a trigonometric identity and rearranging it to find a new equation for the cosine term, then substituting that in, as in Figure 5.4.

Using the trig identity
$$\sin^2 \alpha + \cos^2 \alpha = 1$$
$$\cos^2 \alpha = 1 - \sin^2 \alpha$$
$$\cos \alpha = \sqrt{1 - \sin^2 \alpha}$$

Substituting this value for cos α back into (2)
$$\frac{\sin \alpha}{\sqrt{1 - \sin^2 \alpha}} = \frac{x}{h}$$

Figure 5.4. Using a trigonometric identity to substitute in for the cosine of the angle α.

From there we can rearrange the equation to make our target quantity h the subject of the equation, and then we can substitute in our value for sin α from equation (1), as shown in Figure 5.5. We set $n_0 = 1$ for air.

$$h = \frac{x\sqrt{1 - \sin^2 \alpha}}{\sin \alpha} = \frac{x\sqrt{1 - n_1^2 \sin^2 \theta}}{n_1 \sin \theta}$$

Figure 5.5. Re-arranging to find the height above the water that the beam hits the wall.

Extra things to think about

How would the answer change if you changed just n_1 or θ? What happens at extreme values?

---

[1] You could just as equally well say that we have sin α, but need tan α.
[2] Sometimes referred to as asin or arcsin.